IE. RED T... SANDWICH. ON ... GI... FLU.
PPY BURGER. RUSSIAN ARMY. SINNER'S JUICE. STRAWBERRY
GYPTIAN. BLEEDING BEAUTY. RED SEA. IN DRY DOCK. BLOODY
E OF COURAGE. FUNNY FANNY. MY BLEEDS. FEMALE MOMENT.
DREAD. BAD WEEK. RE... BUNNY WEEK.
S. OLD FRIEND. ON HEA... RIP TO BLOOD
PECIAL FRIEND. MISTRE... SSION. LADY
UTTY. CODE RED. ON THE BLOB. WOMEN'S TIME. THE REDS.
S FIRE. RED FLAG IS FLYING. TIME OF THE MONTH. PAD TIME.
SIT FROM CAPTAIN BLOODSNATCH. PAINTING THE TOWN RED.
ALL. FEELING PERIODICAL. GRANNY PANTS WEEK. HAVING A
FT. BLEEDING CLAM SEASON. MY MENSES. ON FIRE. PAINTERS
ORN. ROLLOVER WEEK. CRIMSON WAVE. STRAWBERRY WEEK.
FT. RED LETTER DAY. A VISIT FROM CAPTAIN BLOODSNATCH.
A PHASE. RED RAG. THE DREAD. TIME OF THE MONTH. BLOODY
PLUG TIME. MY MENSES. PAINTING THE TOWN RED. TRIP TO
P OLD RUSTY. WOMEN'S TIME. LITTLE RED MOUSE WEEK. RED
IDING THE COTTON UNICORN. BUNNY WEEK. POTATO HARVEST.
IVER. WOMEN'S FIRE. SPECIAL FRIEND. DEVIL'S WATERFALL.
L. IN DRY DOCK. FUNNY FANNY. SINNER'S JUICE. AUNTIE FLO.
BLOODY MARY. IN THE RED. RED TRAFFIC LIGHT. MOON TIME.
IS VISITING. GRANNY PANTS WEEK. COMING ON. ON THE BLOB.
ELLIE. BLOOD RAIN. BAD WEEK. CODE ROUGE. KETCHUP WITH
ER. RIDING THE RED TIDE. SURFING THE CRIMSON WAVE. GIRL
OF COURAGE. ROLLOVER WEEK. MY BLEEDS. RED TENT TIME.
S TIME. VAMPI... RED ROSES.
BEAUTY. RED SE... HE DREAD.

Praise for
PERIOD.

'I wish this book had been written before I stopped having them. I might have enjoyed them more! Brilliant, informative and funny. Period.'

JENNIFER SAUNDERS

'I want to hear what Emma Barnett says about everything – and this terrific and timely book proves to be no exception.'

ELIZABETH DAY

'With her uniquely funny and forthright persistence, Barnett de-stigmatises all things period: from the ghastly slang (no, the painters are not in) to the desperate tampon-up-the-sleeve office bog dash.'

PANDORA SYKES

'Why has it taken so long for this powerful, fearless book to be written. It had to be done, and thank goodness Emma Barnett has no shame.'

EMMA FREUD

'A terrific piece of work. Unique. Never read anything like it – polemical, personal, pithy and funny. Sheds so much light on a huge truth that's been standing amongst us forever yet managing to hide in plain sight.'

RICHARD MADELEY

'It doesn't surprise me that Emma has grappled with the "woman's" topic that both sexes have ignored for millennia and made it so completely un-grim, informative and entertaining.'

RACHEL JOHNSON

'A must-read for everyone. Period.'

KIRSTY WARK

'A brilliant, informative, myth-busting, funny and poignant book. I defy readers not to be affected and think differently about menstrual blood – and I defy them not to laugh out loud.'

HELEN PANKHURST

'Clever, useful and wise. Read it. Pass it on to your daughters. And then to your sons. And then read it again yourself.'

FI GLOVER

'Passionate, informed and thought-provoking. How did we let our periods be any sort of secret shame?'

JANE GARVEY

'Emma Barnett cuts right through all the myths and embarrassment with searing facts, honesty and – perhaps most importantly – humour. A bleeding good read that will have you laughing at just how silly the silence around periods is, but also fired up to do something about it.'

YOMI ADEGOKE

'A brilliant, brave rallying shout out to anyone who's suffered in silence and wondered why they had to.'

EMILY MAITLIS

'Smart, funny, enjoyable and tackling a subject head-on that so many others would choose to avoid. I bloody loved it!'

BECKY VARDY

'This book shows what I've always known: Emma's got balls. Or, perhaps not.'

NICK FERRARI

'Brave and fearless, Emma unpacks, demystifies and fights back against the taboo to give us an open, honest and achingly funny celebration of menstruation.'

AMIKA GEORGE

Emma Barnett is an award-winning broadcaster and journalist. By day, she presents *The Emma Barnett Show* on BBC Radio 5 Live in which she interviews key figures shaping our times, from the Prime Minister to those who would very much like to be. By night, she presents the BBC's flagship current affairs programme, *Newsnight*, on BBC Two and hosts *Late Night Woman's Hour* on BBC Radio 4. Emma was named Radio Broadcaster of the Year by The Broadcasting Press Guild for her agenda-setting interviews. Previously, she was the Women's Editor at *The Telegraph*. She now writes a weekly agony aunt column, 'Tough Love' in the *Sunday Times Magazine* and is a proud patron of Smart Works. *Period.* is her first book.

EMMA BARNETT

ONE PLACE. MANY STORIES

HQ
An imprint of HarperCollins*Publishers* Ltd
1 London Bridge Street
London SE1 9GF

This edition 2019

1

First published in Great Britain by
HQ, an imprint of HarperCollins*Publishers* Ltd 2019

A catalogue record for this book is
available from the British Library.

ISBN: 978-0-00-830807-0
TPB: 978-0-00-830808-7

This book is produced from independently certified FSC™ paper
to ensure responsible forest management.

For more information visit: www.harpercollins.co.uk/green

This book is set in 12/16 pt Sabon

Droplet image © Shutterstock.com

Printed and bound in Great Britain by
CPI Group (UK) Ltd, Croydon, CR0 4YY

Contents

For my two boys –
the best team I could wish for

FRESH BLOOD

'Women have been trained to speak softly and carry a lipstick. Those days are over.'

Bella Abzug better known as 'Battling Bella', lawyer, activist and a leader of the US Women's Movement

I loathe my period. Really, I do. I cannot wait for the day it buggers off. For good. But shall I tell you what I loathe even more? Not being able to talk about it. Freely, funnily and honestly. Without women and men wrinkling their noses in disgust as if I'd just pulled my tampon out, swung it in their face and offered it as an hors d'oeuvre.

Don't get me wrong – I am grateful to my period too. A functioning menstrual cycle is, after all, one half of the reason we are all here in the first place and able to procreate, should we wish to. I may loathe the physical experience of my period but that doesn't mean I can't and won't fight for the right to converse about it without fear of embarrassed sniggers.

Periods really do lay serious claim to the label 'final taboo'. But why, in the twenty-first century, are they still seen as disgusting and something a woman should endure peacefully, without fuss? This is despite most other 'off-limits topics' losing their stigmas and coming into the light, helpfully driven by Generation Overshare. But the sight or sound of blood in pants? Don't be daft.

Most women don't even want to talk about them with each other – because there is a deeply rooted idea they are a silent cross to bear, are vile and don't merit anything more than a passing mention.

From their very first bleed, this occurrence in women's pants has been treated by most people around them (female and male) as something to be quietly experienced and hidden away. Periods still have a whiff of Victorian England about them; a stiff upper lip is expected when it comes to what's really going on down below. And women have become so adroit at sparing men's blushes and shaming each other that they have either wittingly or unwittingly denied themselves the chance to talk about their periods, becoming weirdly active participants in the great global hush-up.

Yet, through my journalism and extremely painful personal experience over the last five years, it slowly started to dawn on me that, although on the surface there is a reticence to discuss periods, there's actually a shy hunger to do so underneath, which, when prodded, gives way to some of the most extraordinary tales.

Periods have literally followed me around my whole life. I found myself to be one of the few schoolgirls happy to chat about the red stuff – a taboo I continued to enjoy breaking as I headed into adulthood – much to the chagrin and bemusement of those around me. Little did I know I would become the first person in the UK to announce they were menstruating on live television news; that my undiagnosed period condition – endometriosis – would nearly cost me my chance at motherhood; and that I would be secretly shooting up hormones ahead of one of the biggest political interviews of my career. I hope that on these pages I can bring these narratives together, make some sense of them, and crucially offer some hope, solace and wisdom to women about their periods – served up with a healthy dollop of humour and honesty.

The silence and public attitudes towards periods hold women back – often without us realising it.

Unlike pregnancy and childbirth, periods, the only other bodily process reserved exclusively for women, present no ostensible upside for the male species. Men get nada out of our periods (except, you know, the future of mankind secured).

Plus, if we can't bring ourselves to think our periods merit anything more than a passing lame joke or occasional grumble, it doesn't require a huge leap of imagination to figure out how many men feel about them, if they consider them at all. Horrified. Appalled. Almost insulted. Even the most enlightened man would probably prefer for women to deal with them without breathing a word. And can you blame them? Most of us do everything we can to hide the horror in our knickers, even struggling to talk about periods amongst ourselves.

Men are never going to be the ones leading the charge to stop periods being treated as gross, difficult events. It's down to us women to proudly step out of the shadows and not give two hoots about what men think. It's not going to be easy, and women have to get used to not being everyone's cup of tea. We must ignore the men – and women – who would rather we stayed quiet and 'ladylike' about our periods.

Women censoring themselves from talking about their periods is the final hangover from a time which demanded that we should be seen and not heard; always happy and never complaining; pure and never sullied. It's ludicrous that women remain slightly horrified by something so natural. These women are actively impeding their own ascent to equality with men, for whom nothing is off limits. Yes, women living in Western countries have equality enshrined in the law and yet, we still aren't fully equal. No longer are we confined to a special biblical red tent; we're in your boardrooms, your armoured tanks and we're even running a few countries. But we still aren't equal to men in terms of power, public office and, most damagingly, the way we are perceived.

By allowing periods to remain a taboo, women are imprisoning themselves. Even more worryingly, it's contagious. Girls (and boys) already suffer with low self-esteem and that's only getting worse in the social media era. When it comes to simple bodily functions, the least we can do is remove a stigma that has damaging consequences on half of the world's population. Many women already judge themselves to be less than men or suffer with imposter syndrome. If we then conceal something that happens one week of *every* month (often longer) we are unconsciously turning our periods into a form of disability, as well as failing to confront negative myths and damaging how we view ourselves.

The bogus presumptions about menstruating women are tragically still not confined to the history books – namely that we are weak, dirty, unhinged, less than and just different. At the heart of this lingering stigma is the idea that we are unequal to our male counterparts. Women then ingest these views and appropriate them as our own, inflicting wounds on ourselves and other women around us. And by keeping periods unmentionable, women become unwitting accomplices in perpetuating these myths.

No more I tell thee, dear reader.

Period shame has been stubbornly hanging around since the beginning of womankind and it's about bloody time for change.

Because if there's one thing we do know, it's that a period waits for no woman, so let's finally allow the period pride to flow.

To be clear, I am not saying women feeling better and bolder about their periods is a secret key to unlocking the door to full equality – if only. It won't stop at least two women a week dying at the hands of a man in the UK, or suddenly catapult a woman into the White House (as the incumbent, and not the wife who picks out new china and curtains). I'm not naive; nor do I wish to overplay one element of our lives.

But the way periods stubbornly remain taboo, along with all other things we hide with shame and fear, is highly symptomatic of how women have been indoctrinated to believe that a perfectly natural bodily function is totally abnormal. It is this attitude, which too many women, men, religious figures and tampon companies propagate, that ensures we remain ashamed of one of the fundamental signs of womanhood at all ages and stages of life.

Women fear being seen as weak in the workplace, so say nothing about menstruation and any issues they might be having, tacitly reinforcing a view that we are less capable during our time of the month.

Schoolgirls in Britain are missing out on their education because their families cannot afford to buy tampons or pads. Period shame stops them asking for help or admitting why they are skipping school. The same is true of fully grown women who can't afford pads. They stay in their homes, imprisoned by the fear of someone noticing they are unwillingly leaking through their makeshift sanitary pad (sometimes it's a sock, other times, loo roll). Nobody knows so nobody helps. A totally unnecessary cycle of period poverty remains unbroken and wreaks havoc. In the UK, the world's fifth largest economy. Right now.

If we don't start talking openly about these issues, these perceptions will go unchallenged for yet another generation. Periods should be as natural and as unremarkable as waking up with a headache or needing to pee. And until they are, women and girls will remain relegated – different and unequal within our very selves.

Simply put, periods shouldn't be seen as a source of shame. Instead a period should be seen as a sign of health, potential fecundity, strength and general bad-assery.

Don't be revolted, lead the revolt – preferably with a grin on your face and a tampon tucked proudly behind your ear.

There's nothing else for it. You didn't realise it and neither did I until recently – but we need a period crusade. For our health, our happiness, and because this bizarre taboo is holding women back.

This book, and the stories within it, is for all women, and not just for the minority who are already comfortable shouting about every part of their existence. But it's also for men who want to understand what's really going on in the lives of the women and girls they care about.

My aim is to normalise every aspect of periods, to mention the unmentionable and help you notice things you've never considered before. And crucially, to make you laugh along the way. Women, it turns out, can be extremely funny about blood. Periods can be a subject worthy of mirth. Who'd have thought it?

I spotted my personal favourite comment on menstruation upon an e-card, which read: 'Why periods? Why can't Mother Nature just text me and be like, "Whaddup, girl? You ain't pregnant. Have a great week. Talk to ya next month."'

Ultimately, I'd love to instil within you, my wise, merry readers, a sense of period pride, perspective and some flipping normalcy around menstruation. Because unless we change the way we talk about periods, this silence and shame is here to stay.

So, as my Eastern European pal used to sardonically say each month: the Red Army has arrived.

But the big question is: are you with me?

CHAPTER ONE

'Girls are superheroes. Who else could bleed for a week and not die?'

(A very true internet meme)

All women remember their first period. Where it happened; who they were with; what raced through their mind and what they did about it. Or didn't. The sight of blood anywhere is frightening. In your pants, it's terrifying.

I want to share the story of my first period with you for two reasons. One – quite frankly it would feel rude not to in a book about the red stuff. Two – my mum's reaction goes some way to explaining why period pride came quite naturally to me.

But even though I have felt pretty confident about busting the period taboo at each stage of my life, to my horror I still utterly failed to achieve a diagnosis for a serious period condition ravaging my insides for more than two decades. I wish to open up to you about this particularly agonising chapter of my life to show that even if you have never spoken about periods aloud before – yours or someone else's – you can start now. And you should.

Apart from kicking off a much needed cultural shift around the silence that engulfs periods, it was only when I admitted to a friend how much pain I was in each month, that she suggested I might have a proper illness, which prompted me to push for a GP referral to a specialist. More women need to be heard to be believed. And only by talking more about our periods, can we learn what's normal and what isn't – and that actually, we're bleeding superheroes.

I started my period just shy of my eleventh birthday in a cold toilet cubicle in Manchester's House of Fraser. As I was an only child with a devoted mother, who delighted in my every milestone, I shouted out to her from the cubicle that there was something browny-red in my knickers. She told me, breathlessly, that I had indeed 'become a woman' and started my period. I'd just started reading Judy Blume books and had a vague idea this was a good but major thing. And then she left. In a panic. Off she ran around the whole shop floor telling anyone and everyone her little girl was having her first bleed and asking around for a spare pad. Subtle. A few minutes later, I opened the door and watched as she gently stuck a large pad down into my stained pants. Her excitement was infectious.

I remember feeling like I'd done something positive and exceedingly grown up. And when we walked out of the loo, I recall bashfully taking in the smiling faces of the female shop floor staff, as if I'd just won gold in the Woman Olympics. I now know my lovely mum was trying to make up for how her own mother had reacted to her first period. My mother was told, in the swinging sixties no less, that she was ill and put to bed. No explanation was given, but it became clear that it was a subject that was off limits for discussion – the final irony being that her father was a doctor! It was a terribly confusing, scary and negative experience for her.

Mine couldn't have been more different. On the day of my first period – over a celebratory steaming hot chocolate – as my mum delivered a basic explanation of what had just occurred (something along the lines of 'this will happen every month and welcome to the woman club'), she was beaming and almost crying with pride. I was excited, but I also remember asking her not to tell my father. I'm not sure why I wanted to keep it from him but it was probably because it concerned something dirty in my pants to do with my vagina and he didn't have one.

Without even realising it, I was already hard-wired to protect the man in my life from potential female grossness.

Either way, when we got home, she swiftly reneged on our agreement. In fact, we hadn't even taken our coats off before she proudly told him that I had Become a Woman. Any feelings of anger I had at her big mouth were swiftly cancelled out by his understated but lovely response. Probably a touch confused as to what to say, he sweetly wished me *mazel tov* (like you do as a northern Jewish father who doesn't speak that much), and swiftly went back to reading his *Manchester Evening News*. And that was that.

Except it wasn't. Life had changed and my first period lasted for nearly three weeks. I don't remember the pain with that one – that was to come later and to define my whole experience of periods. But the discomfort of large nappy style pads in those long first three weeks still looms large in the mind.

I dimly recall telling a few friends at school 'I had started', but I was one of the first to get my flow, so it was only a small number of girls who knew what I was talking about. It was to take one of my closest friends a further four years to get her period so she was oddly quite jealous. Consequently, it didn't feel right to complain to her about my need to shift my heavy knickers about in a distinctly unladylike way in class, trying to get my new massive nappy into a comfy position and stop the adhesive underside from sticking to my legs. Mum hadn't really catered for small knickers in her choice of sani pad.

My main period chat was with my mum during this first flow, as she inspected my pad and asked how I had felt before and after school every day of my biblically long bleed. And because of this regular checking in (during which my doctor grandfather was also consulted on the phone as I entered my third week, and I was breezily assured all was well), my first experience, unlike so many women's, felt OK. Cool even. Mum and I made some daft jokes and I definitely got some extra chocolate. Periods were made to feel like a new inconvenience, but one I could totally handle. Yet so many girls suffer in ignorance and silence during their first period, establishing an embarrassment they carry for the rest of their lives.

My period was also on the conversational menu at school if I wanted it to be. My girls-only school – Manchester High School for Girls, where Emmeline Pankhurst sent her strong-willed daughters – may not have given any of us the full biological low down until we were a bit older, but the largely female staff were always receptive to chats about the red stuff. (Especially the sceptical swimming teachers who listened to our tales of gore, fake and real, as we soon learned the quickest way to dodge the icy school pool was by saying we 'had a really heavy period'. Finally, a benefit!)

Soon, though, this was a reality for me. After my relatively pain-free maiden bleed, my period quickly became a much darker experience. Clots and crippling cramps were my new norm and I found myself wincing in pain for the first two days of every cycle. My mum, who I soon learned also suffered painful heavy bleeds and was upset that history was repeating itself, was swiftly on the case.

Interestingly, I felt I could openly confide my fears to my mum but I didn't feel I could, or should, talk about it with most of my friends – and not just because some girls were jealous, I now realise. Whilst I was prescribed contact lenses at a similar age, it felt like something I could openly bitch and moan about, no matter how insecure I felt. My period, not so much. Even when I rushed to the school loos in crippling pain, and tried to explain why to my peers, their blushing cheeks meant that I was already being socialised to keep quiet about my period. If this is what it's like in a girls-only environment, is it any wonder that we've all stayed silent for so long? And I didn't even get teased by horny pre-pubescent boys about smelling or being frigid because I was 'on the blob' – unlike the experiences of so many girls I know who had brothers or went to mixed schools.

Around my twelfth birthday, when the monthly pains really set in, I spoke up again to my mum, refusing to believe my debilitating experience was normal and was swiftly taken to the GP who prescribed strong period-specific painkillers called mefenamic acid. I have vivid memories of both parents soothing me, as I writhed around in my bed and found myself with the runs, hobbling to the loo (yet another side-effect of periods that remains a taboo subject).

The point is, I *had* the period confidence to do this and suffer openly (at home at least).

I was made to feel proud on the day I bled for the first time, rather than dirty and ashamed.

Periods weren't taboo with the few adults in my immediate daily life. I realise now, in my early thirties, how enlightened and important that reaction was. Another friend's parents broke out into a congratulatory song in front of her brother and his pal, when she started. While she was mortified, just as I would be (not because of the subject but actually because I loathe spontaneous group singing) it also instilled within her a sense of happiness and achievement on her first period day.

I was always encouraged to talk about my feelings – something which has stood me in good stead for becoming a broadcaster. But that doesn't mean my friends and I fully understood our periods, as evidenced by our rather disastrous attempt to help a girl insert her first tampon aged sixteen on a school trip. More of that to come . . .

However, the open attitudes both at home and in my schooling, and being encouraged from an early age to challenge boys at every opportunity – especially on the school bus when 'banter' was at the girls' expense and was often regarding matters of puberty – led to me possessing something so few women and girls have: period pride.

I mentioned it in the previous chapter. But stop for a moment to consider the phrase. It consists of two words you don't often associate with one another – let alone see written down together. And that's what I would really like to inspire in you. I want to infect more women with period pride and, in turn, cure men of their need to retch when the topic arises.

Period pride doesn't mean you have to enjoy your period. I certainly never do. My periods have been defined by bone-grinding pain; I have never been one of those women who breezed through their monthly bleed. It wasn't something I felt I could or should ever ignore.

At school, I didn't talk about my period much, though I found it odd that no one else was screaming about this strange monthly occurrence. The only place I found any more information about it was within the comforting pages of Judy Blume's *Are You There, God? It's Me, Margaret*, in which the protagonist begs God for her period to come so she can be normal. Incidentally, this tome is one of the most challenged and banned books from US children's libraries – make of that what you will. 'Put up and shut up,' was the vibe around periods.

Once our paltry sex education kicked in when I was around 13 or 14, they were only spoken of in the truest biological sense. No mention was made of our hormones, moods or other parts of our bodies and mind. The tone was also rather glum, portraying it all as a bloody cross to bear rather than something faintly ludicrous, or an experience to exalt in any way. Had just one teacher broken ranks to tell us her worst leak story, we might have at least enjoyed a giggle at our shared shedding of innards.

Unfortunately, there was not that kind of female solidarity at my school, nor really thereafter, but that didn't dent my natural period pride. In fact, I became even less meek about occasionally busting out some of my period's greatest hits to my appalled male and female university housemates and, later, to my closest pals and colleagues. Even now, my husband, despite often being a better feminist than me, still occasionally wrinkles his nose in squeamish disgust.

'I'M MENSTRUATING. RIGHT NOW. ON YOUR TV.'

My openness about my period is why I found it uncontroversial, perhaps weirdly easy in fact, when I made a spot of broadcasting history on 19 May 2016. You see, I am the first person, certainly in the UK, to look down the barrel of a camera lens on a popular 24-hour news channel and unashamedly utter the words: 'I'm menstruating, right now.'

The TV programme? *The Pledge*, a free-talking evening panel debate show I used to co-present on Sky News. The reason? I wanted to debate menstrual leave in the workplace. And the panel? Duly horrified.

The subject was in the news that week because a small Bristol-based company, aptly called Coexist, had become the first in the country to introduce a 'period policy' in a bid to help its large female workforce be more productive. Practically, this meant employees could take time away from the office around their menstrual cycles and work more flexibly at that time of the month, instead of – as the boss put it – being hunched over their desks in pain.

Menstrual leave is a common policy across large parts of Asia but has yet to catch on as a Western phenomenon and I had very mixed views about it. While I applauded the concept of breaking the taboo around periods, and giving women the option to work more flexibly during heavy or difficult bleeds, I also loathed the idea of periods being weaponised against women as a badge of weakness in the workplace. Or, that it could become a fashionable policy, writing off our periods like some politically-correct fad, something that is simply a 'hot topic of the moment' only then to be forgotten and glossed over.

I soon realised I could use my lack of squeamishness about my own periods to great effect. I had the chance to kick off a real debate and duly wrote my autocue script with a wry smile on my face, knowing it would get a reaction. My wonderful producer, always encouraging of anything that would enliven our conversations, was thrilled, almost titillated. I was happily putting myself out there, the subject was fresh, and she knew it would expose divisions in our diverse panel about a genuinely controversial issue.

Usually we rehearsed our opening statements in front of each other before recording, but for this topic, we decided I should do it on my own, to keep it fresh. The hilarious thing was that everyone apart from me – the camera crew, the producers in the gallery and the editor – were jittery and on tenterhooks about my big reveal. Especially as one of my co-hosts was my faux adversary, the masterful broadcaster Nick Ferrari. Off air, I love Nick and he is a very good pal, but on air we pretty much disagree about everything, especially gender issues. It was a subject that had the potential to generate fireworks and Nick certainly didn't disappoint. From the minute, I began describing my painful clotting flow, he slowly lowered his head into his hands, muttering: 'Why, oh, why did we need to talk about this, Emma?'

His cheeks were bright red and he looked like he might vomit.

He wasn't alone. Both my female and male colleagues – including journalist Rachel Johnson and international footballer Graeme Le Saux – were howling with shock and dismay. Intelligent adults *howling* on national TV. They didn't know whether to laugh or cry. But, because I'm a good humoured kinda' gal, wanted a proper debate and loved my fellow Pledgers, I swiftly helped my colleagues through the shock of 'Woman Admits Period on Live TV' and attempted to move them on to the issues at hand.

The irony was of course that, by attempting to lead by example and not be ashamed of talking about my own period, we never really fleshed out the debate. I was genuinely trying to start a conversation about whether menstrual leave was an appropriate response by company bosses, or whether it was a policy ripe for enforcing the age-old idea of women as the weaker, less capable sex. Instead, the laughter-filled conversation became more about why I felt it was necessary to make a big deal of periods. Put simply, my co-panellists couldn't make it past the fact that I'd mentioned my own pulsating uterus on TV.

But what was even more striking was that the reaction of my female panellists was just as strong as the men – they too argued that this cone of silence should continue. I ended up feeling terribly alone, as their shock – and I believe, embarrassment – drove them to giggle and sneer about the whole discussion. Trust me, this isn't a group who shock easily, but they couldn't get over the fact that periods were being discussed out loud, on TV. And when other women aren't setting the tone for men to follow, it's hardly a surprise that my male co-hosts couldn't bring themselves to go beyond some gentle scoffing and a few end-of-the-pier jokes.

Interestingly, once we were off air, one of the lovely female panellists said to me in the make-up room: 'Perhaps I should have been a bit more supportive of you out there on second thoughts, but there we go.' Rachel Johnson, one of the most open and outspoken journalists I know, had turned period pink with an uncharacteristic bashfulness, over something so natural.

Several weeks after the programme aired, I was thrilled to hear that one of the women in the editing gallery (often a male-dominated space) was suffering from particularly painful cramps, and when asked if she was OK, gruffly replied, 'I've got what Emma talked about'.

I was also cheered by the reaction of viewers at home, with messages of support for my efforts pouring in on social media. Men revealed that their female partners never spoke to their colleagues about their discomfort. Women thanked me for my candour, realising how rare this kind of honest conversation was – and their response has stayed with me ever since.

A few months later, a woman stopped me in my local fruit and vegetable market, clutching at my arm with fervour that was slightly alarming. She was a lawyer and told me that she'd seen *The Pledge* episode – it had been a lightbulb moment for her. Close to tears, she recounted how her heavy period had flooded on the bus on the way to work one time. Ashamed and panicked, she'd desperately scrubbed and dried her suit in a coffee shop loo in case her colleagues noticed. Sadly, it wasn't an isolated incident and she was diagnosed with fibroids – her monthly downpours forced her to quit her full-time job, turning her into a part-time shift-worker so that she could manage her working life around her cycle. She'd never spoken about this to anyone and wanted to thank me for speaking out about periods without shame – it was this that had really struck her, how simply I had admitted I was menstruating and in pain on national TV.

It was like I had started to unleash something people hadn't known needed releasing: the shy hunger to talk about periods. It dawned on me that enabling others find their voice on periods and engaging in this forbidden conversation could be an important step in helping women stop judging and shaming themselves in all walks of life. Stories of pain, shame and hilarity have followed me around ever since.

Sitting in that glamorous, shiny-floored studio, what I didn't realise was that I was about to be diagnosed with endometriosis – a common but often poorly diagnosed menstrual condition (where tissue similar to the lining of the womb starts to grow in other places, such as the ovaries and fallopian tubes) that causes millions, yes, *millions*, of women colossal amounts of bone-grinding pain every month and can have serious consequences for fertility. It turns out those painful early periods I'd had weren't normal or acceptable. Me, the one with the big mouth and period pride, actually had a proper period illness all along. One I'd never heard of, couldn't spell nor explain.

I soon underwent the most painful forty-eight hours of my life, as a surgeon lasered my insides for nearly three hours. Although it's estimated at least one in ten women in the UK have endo (as it's called by those in the painful know), it takes a scandalous seven and half years on average to be diagnosed with this progressive illness. It took me nearly twenty-one years to get my diagnosis – and even that came down to pure luck.

By chance, I'd opened up about the pain I was in every month to a doctor mate over Sunday brunch, and she'd tentatively suggested that endometriosis might be a possibility. I'd had over two decades of medical appointments in which I complained repeatedly of severe menstrual pain, but I'd still failed to convince my doctors to take my period pain seriously. I'm someone who thrives on bashing down doors and demands answers for a living, yet I still hadn't got anywhere – so what hope was there for others?

To be frank, I am fucking furious that nobody knows what causes endometriosis – mainly because it's a 'woman problem' and there hasn't been enough investment in scientific research. And people, even gobby folk like me, walk around ignorantly ill because no one takes our complaints of abnormal pain seriously, so we just stop talking, afraid of looking like we're moaning.

As I hobbled around for three months recovering from my debilitating op, I felt as if the rug had been pulled from beneath me. But it meant that periods were back at the top of my personal news agenda, and I felt bloody foolish.

Back at work and intrigued by the silence and ignorance surrounding our periods, my team then commissioned a study for my radio show. Over 57 per cent of women who told us they had period pain (which was 91 per cent of the total number of women polled), admitted it affected their ability to work. But only 27 per cent of them told their employer the *real* reason they felt poorly, with most preferring to lie, often opting to say they had stomach problems. Interestingly, many British women were open to the idea of menstrual leave.

Before presenting the findings in our programme that day, I was invited onto BBC *Breakfast* to discuss the results – an invitation I happily accepted. Here's how *The Sun* decided to write up my calm and measured conversation with my BBC colleagues: 'Woman Sparks Furious Debate About Menstrual Leave on BBC *Breakfast* – BBC Radio 5 Live's Emma Barnett Has Sparked a Sexism Row.'

Now who's hysterical eh? And people still wonder why women lie about their periods, preferring to tell their bosses (of both genders) they have the shits. Go figure.

Periods *need* to come out of the darkness because of the potential benefits to women's health around the world. But the cultural benefits of smashing the period taboo would be major and just as important.

Alisha Coleman, an American 911 phone operator, sued for alleged workplace discrimination after being sacked for leaking during two particularly heavy periods. Once, on her seat, for which she received a written warning, and then again on the office carpet. Can you imagine how mortified she felt? How you would feel? Perhaps you've been there too. Alisha was dismissed for failing to 'practise high standards of personal hygiene and maintain a clean, neat appearance whilst on duty' the lawsuit states. I am utterly dismayed that women can lose jobs over leaking on their office chairs.

'I loved my job at the 911 call centre because I got to help people,' Alisha, a mother of two, explained. 'Every woman dreads getting period symptoms when they're not expecting them, but I never thought I could be fired for it. Getting fired for an accidental period leak was humiliating. I don't want any woman to have to go through what I did, so I'm fighting back.'

Quite.

But Alisha isn't alone in struggling to conceal her period. One woman confessed to me, in the bespoke period confession booth built for my BBC 5 Live radio show, that she once had to call in sick because of severe period pain. She decided to the bite the bullet and be honest when she spoke to the HR director (who happened to be a man): 'I could hear him on the phone being squeamish [when I told him the reason] and then he said, "Well let me know when you are fixed."' *Fixed*. As if she was a broken doll. I interviewed six other women that day, in their twenties, thirties, forties and fifties, and they all had their own coping mechanisms for making sure their period experiences remained as hush-hush as possible. You can understand why.

The very fact that my production team, who happened to be all women that day, had been delighted to build a beautiful confession booth (complete with a gold grill), spoke volumes about the need to have the conversation. The storytelling power of radio may be in its facelessness, which emboldens even the meekest guest, but we still felt the need to create a further physical divide to coax these successful and confident women (with the strict assurance of anonymity) into the box to talk about something as natural as their periods. Upon reflection, it was utterly ludicrous.

This censorship and secrecy that engulfs periods was highlighted by the global shock, teetering on outrage, when Kiran Gandhi decided to run the 2015 London Marathon free bleeding down her Lycra leggings. Kiran came on her period the night before and, deciding it would be uncomfortable to run 26 miles with a stick of cotton wool wedged up her genitals (you think?), chose to let it flow. Admirably, she also did it to raise awareness for poorer girls and women who don't have access to period products. Whether it did that or not, I cannot say. But I do know many newspaper picture editors will have furrowed their brows as to whether they could run such a 'distasteful' image – despite it being all over the internet and the very same news outlets thinking nothing of filling their daily feeds with graphic images of bloody wars. You will hear more from Kiran a little later on.

Does that not strike you as terribly odd, now you stop to consider it? The fact that the sight of a woman's menstrual blood, coming out of the hole it's meant to, provokes more consternation than the image of lifeless children's bodies being pulled out of the Mediterranean Sea in their failed fight to find a permanent home? One image can be published without question, while the other is censored, out of fear of poor taste and offence.

In China, tampons are still thought of as sexual

items rather than basic sanitary goods. Caught short on holiday, when I asked a young female shop assistant in Beijing where the tampons were (whilst making unedifying finger gestures as I stood next to rows of sanitary pads), she looked at me like I was a strange slut.

There is still a fear, especially amongst younger Chinese girls, that tampons break the hymen. Recent figures show that only 2 per cent of the country's women opt for tampons, compared to Europe's 70 per cent. According to my friend and fellow journalist, Yuan Ren, Chinese medicine is also to blame for the cultural fear of tampons. It propagates the idea that putting a foreign object into the body is not good for you and can be harmful to girls who are still growing. Add this to the lack of education in the country as to how to insert a tampon and you have the perfect combination of factors to ensure that millions of women never experience the joy of a less cumbersome absorption method.

Instead, they have to waddle about with a cotton surfboard jammed between their legs. Lovely.

So why don't we just say it how it really is? Periods are the blood of life and in many ways go to the very heart of being a woman. Simone de Beauvoir, when writing her seminal feminist text, *The Second Sex*, in 1949 wanted us to have jobs and work through them. She believed distraction from our pain was a good thing and the only way to cope. Germaine Greer, in her similarly iconic 1970 bestseller *The Female Eunuch*, famously wanted us to taste them as a test of our emancipation (full disclosure – I haven't). She wrote: 'If you think you are emancipated, you might consider the idea of tasting your own menstrual blood – if it makes you sick, you've a long way to go, baby.'

Periods, whether you are into tasting yours or ignoring it as best you can, are part of the essence of being female.

And yet we vilify them, disproportionately.

We know that men, particularly recent American presidents, have difficulty with them. But women also hush up other women on the topic, mainly out of fear that they will be seen as weak. Just because periods come out of our nether regions and, by their nature, are messy things, women worry this renders them worthy of censorship. But why? Is there just a horror and shame linked to blood in the pants? Granted, many men wouldn't want to talk about their bleeding haemorrhoids. But they aren't a normal monthly part of life and a vital sign of health. Quite the opposite. Or is it something that runs deeper and proves that women have bought into the age-old myth that anything uniquely female is filthy, reductive and not quite right. That we are broken and yucky in some way.

We can't continue as a human race without periods – and yet we still can't acknowledge their existence. In the twenty-first century. I am not calling for women to walk down the street in short skirts with their tampon strings dangling out, armed with megaphones screaming: 'Look at us! We're bleeding!' (Although if you want to do that – go for your life, sister.) But what I *do* want is for this juvenile shaming attitude towards women and a vital part of our anatomy and health to stop being such an embarrassing mysterious and dirty secret.

Most women I know wouldn't walk to the toilet in their office – a place they go every single day – without a dainty 'special zip-up bag' (ladies, you know what I'm talking about here) or even their whole handbag, just to take a tiny tampon into the loo for a change. Can you imagine if men bled for a week every month? Some form of menstrual leave would have been written into HR policies around the world, period-pain-bragging would be an Olympic sport and bleeding males would dramatically stagger to the office bog with their tampon proudly gripped in their fist. There would be no need to hide sanitary products in tiny zip-up bags. But men don't have periods. And history has meant that *they* were the ones who designed society and the world of work without women – or our monthly downpours – in mind.

It's time to perfect your period patter and swagger with pride, but it's also important to know what you are up against: generations and generations of debilitating myths and anti-women, fear mongering nonsense.

Just as I began this chapter, that's why it's important to remember that 'Girls are superheroes. Who else could bleed for a week and not die?'. I've got this. You've got this. We've got this.

Turn this page, turn a new leaf. Our work begins now.

CHAPTER TWO

HOLY BLOOD

'The most common way people give up their power is by thinking they don't have any.'

Alice Walker, author and poet

Before I tell you how periods unexpectedly took centre stage in the run up to my wedding, I thought it first prudent to share a list of some of the codswallop that women on their period have been blamed for and prohibited from doing while menstruating.

Menstruating women must not:

- Make mayonnaise as it will curdle
- Come into contact with mirrors as they will cause them to dim
- Walk through fields of courgettes, pumpkins or fruit trees – all will rot and wither
- Contaminate butter as it won't churn
- Touch wine as it will turn to vinegar

- Venture near dogs as they go crazy in the vicinity of period blood
- Go camping or wild swimming – bears and sharks are drawn to the bleeding lady
- Walk in front of anyone as their teeth will instantly break

Believe it or not, some of these nonsensical outright lies are still doing the rounds today. The teeth-breaking one? Still frighteningly alive and well in modern-day Malawi. The mayo gem? Still very much in rude health in Madagascar.

While we'd like to believe many of these myths – religious or otherwise – have died a death (along with believing women aren't capable of doing loud smelly farts or pumping out logical thought), stories have a way of burrowing invisible roots deep into society. Even if there's no basis for the prejudice or superstition, a grain of belief can still linger for a long time afterwards, colouring people's views. So, although no doctors in the UK today believe that menstruating women could spoil meat, as some writing in the respectable *British Medical Journal* in 1878 did, there's still a hangover from that type of 'intellectual' discussion which saw women as being 'dirty' during their period.

It is clear that religions haven't been solely responsible for all period myths – doctors, tribe chiefs and the great thinkers of the day have all contributed their own bits of gibberish. However, all of the major faiths do still have a lot to answer for, as I found out to my surprise the year I got married.

I was born into a Jewish family, and brought up culturally Jewish – so, big Friday night dinners, a decent level of Jewish education until the age of twelve at Sunday school and attending my fair share of Bar Mitzvahs. And despite not being particularly religious or observant, the ideal romantic situation envisaged by my family was that I would eventually find a Jewish guy.

I always explain to people who struggle to understand why you might prefer to marry Jewish if you are Jewish but aren't that religious, that it's akin to wanting to marry someone from a similar background to you. That's all. Someone who immediately gets your weird home rituals without explanation, understands your family's quirks and with whom you have a shared history. But I should stress I probably would have also married outside of my faith too – because I believe in falling in love which is nigh on impossible to prescribe.

On a practical level, it's really tough to find a Jewish mate, especially in the UK where there are now fewer than 250,000 of us in total, and the only part of the community which is growing in number is the ultra-orthodox. So, finding a Jew who is similar to you in terms of religiousness and outlook (as well as the million other ingredients that go into being compatible with someone) is tricky, especially as you're shopping in a very small store. But, somehow, I did indeed land my match and amazingly, he happened to be Jewish. A lucky bonus for me.

When I met my husband, aged 20, I was wearing a blue Nottingham uni theatre T-shirt with my name emblazoned across the back (sexy, I know), because I'd recently been elected president and, before the journalism bug hit, I harboured dreams of acting and he was wearing stripy Birkenstocks. Also très sexy. I was in a flap and attempting to deal with a severe budget cut to the theatre's meagre pot. Except my grasp of general maths, spreadsheets and deficits weren't the greatest.

My fun-loving mate, Gemma, from my politics class I occasionally attended, had told me that her friend could help – plus, he was single, good at maths and HOT. Boldly, I introduced myself to him, and after some sexy budget chat in front of the theatre's noticeboard I found myself complimenting his Birkenstocks and asking for his number, sober, in the cold light of day.

Long after that first encounter, he told me how bowled over he was by this forward northern woman demanding his digits. Fast forward through many dates, holidays, jobs and postal addresses, we are about to celebrate fourteen years together. And, even though it was daunting having met each other so young, at the peak of sowing our wild oats, we have stood the test of the time (even if the Birkenstocks haven't). But why am I telling you how I met my husband? Because seven years on from that first meeting in front of the noticeboard, we were back there and something he did inadvertently led to us getting up close and personal with my period.

I'd been invited back to Nottingham University to give a lecture to politics students about how to get into the media. My other half had merrily tagged along. It was our first weekend back in the city since we graduated, and a little tipsy on red wine after a cosy dinner, I unwittingly set up my own wedding proposal. 'Wouldn't it be fun to stand in front of the noticeboard on the exact spot where we met?' I asked, excitedly half running to the very point, with him smiling and walking behind me. Five minutes later, my then boyfriend was down on one knee asking me to marry him.

We decided to get married at a synagogue we'd recently discovered in London's Bayswater, while renting locally. We had passed this beautiful building countless times, but being rather rubbish Jews had wrongly assumed it was a church. Finally, having made it inside on a random Saturday and been proven wrong, we fell in love with this Moorish-style temple and were charmed by the friendly local community and the brilliant rabbi, who was modern and amenable to our needs and religious crapness (my words, not his).

Someone in the community mentioned there were people who would happily give us the low down on Jewish marriage if we wanted to hear more about the experience. Always a sucker for learning and the chance to ask questions, I signed us up.

Now, if you were offered the chance to hear more about your faith and marriage ahead of your pending nuptials – what would you expect to learn? Perhaps some wisdom about love, sex, the wedding ceremony, family and being a single unit. I was hoping for tales of love in the Bible and to find out any kosher kissing tips

(I jest. Slightly). My fiancé was just hoping to survive the experience. What percentage of that conversation would you expect to be about periods – a topic you'd never even really discussed at length or in any serious detail with your husband? 2 per cent, if that?

Well, 75 per cent of our informal session was about periods. *My* period to be precise. And how 'impure' it made me for nearly half of every month.

And that's how my husband's romantic university proposal ended up leading to one of the most memorable conversations I've had about my menstrual flow.

We barely had a chance to sit down upon meeting our informal guides, before the foreign concept of *niddah* was brought up.

Before my fiancé and I could exchange quizzical looks, I was invited to speak privately with the female volunteer. Finally, I thought, this was more like it. It was time for the good stuff in my girls-only chat.

Settling into a comfy sofa, the friendly woman began what has now become known in my friendship circle as 'the legendary period talk'. Smiling at me, she said something along the lines of: 'Emma, when you bleed each month, you become *niddah*. Impure. Unclean for your husband. And this lasts until the very last drop of blood has come out of you and you have cleansed your whole self in the *mikveh* pool. Do you understand?'

She then told me that, during this two-week window of time,

I was wasn't even allowed to touch my husband's sleeve, or, in my favourite example, pass him a piece of steak I'd cooked for his dinner.

As I was still digesting her words and mulling over how he was better at cooking steak than me, she began confiding the romantic and practical benefits of *niddah*. She told me that, like with anything in life, restraint makes something sweeter when you have it again after a while. Not touching for nearly two weeks every month meant you couldn't wait to touch each other again, after the *mikveh*. And, handily, this would also be the right time in your cycle to get pregnant. *Who'da thunk it?* Plus, she confided, it's sometimes nice to have a break from sex and your husband for half a month, every month.

My softly-spoken guide sat back, pleased with her explanation of how the Orthodox Jewish way had thought of everything. And while a small part of it seemed plausible (i.e. the part about abstinence making the sexual pull stronger), I felt as if I'd fallen down Alice's rabbit hole and was struggling to re-emerge from Wonderland.

But the truly jaw-dropping revelation was yet to

come. Before entering the *mikveh* pool – a pool which, I should add, you cannot enter with nail varnish on or even your hair plaited (a place my mother had religiously avoided her whole life) – I had to be completely sure that my period had finished. So how can you be 100 per cent sure, beyond your own eyes telling you that your tampon is clear and your pants are pristine?

A kosher rag.

Yes, you read that right. There is a special cloth you can buy to wipe yourself with so that you can double and triple check that your period has properly ended. But *oh, no, no* – that's not all you can do to ensure your purity...

If at the end of your cycle, after you've wiped yourself with said specially blessed rag, you are still in doubt, you can *post* the scrap to your local rabbi in an envelope with your mobile number enclosed. After he's inspected it, you will simply receive a text telling you whether you are kosher or not for swim time at the *mikveh*. According to my smiling guide, they are 'specially trained'.

Yes, Jewish women are wiping themselves with and then posting bits of cloth to a bearded man down the road.

My own brazenness started to falter here, and I didn't press any further. I have often wondered since about the identity of the first dude who figured he had the expertise to pass on this knowledge to all the other men who have never menstruated in their lives. I can't recall much more from that session other than the moment my fiancé suddenly reappeared from his part of the house and swept me out of there. We both felt a bit sick – him from too many salty and sweet snacks, and me from a very odd period chat.

'*They do WHAT*?' was his general reaction, as I explained about the rabbi rag watch.

Meanwhile, in the other room, my fiancé had been given a more pared down version of how periods affect women, with a similar emphasis on 'no touching' during my 'impure' period. But I'll never forget what else he had been told: 'Women are a little crazy when they are on their periods' – with the implication that, in fact, it was good for all husbands to have a break from their wives at this point in our monthly cycle. Yes, really.

(Later, in a totally unrelated conversation with a friend, I also discovered that women who are trying to fall pregnant can post vials of their period blood to clinics in Greece, who claim they can analyse the sample to spot potential fertility problems. It makes you wonder what other bizarre packages are being sent through the post, doesn't it? Again, another rabbit hole.)

I must stress this: the people we met were kind and only trying to educate us in the ways of ultra-Orthodox Judaism. I don't wish to be uncharitable towards them; my criticisms aren't personal. And, of course, other folk could have sold this purity period-obsessed side of marriage to us in a softer way. Or not focused on it as much. But I am grateful for receiving an unvarnished insight into what one of the oldest religions in the world teaches couples about women, our bodies and purpose on this earth.

I know that some modern Orthodox Jewish women have reclaimed the *mikveh* as an empowering space for them to feel cleansed and almost reborn each month – but why is period purity even still a thing in the first place? A step forward in the dark isn't a step forward to me. Teaching girls and women that they are dirty and in need of rebirth after something as perfectly natural as a period isn't right, however it is spun.

Nor am I looking to solely point the figure at Judaism. God knows (and he really does), it's a religion with enough haters already. Judaism is certainly not unique when it comes to the major world faiths pillorying or discriminating against women for menstruating. Far from it.

Factions of Islam believe women shouldn't touch the Qur'an, pray or have sexual intercourse with their husbands while menstruating. Muslim women are similarly deemed impure and must be limited in terms of contaminating their faith or their men.

Catholics fare no better, but seem to prefer to whitewash the whole affair. According to Elissa Stein and Susan Kim's book, *Flow: The Cultural Story of Menstruation*, when Pope Benedict visited Poland in 2006, TV bosses banned tampon adverts from the airwaves for the duration of his stay – in case his papal Excellency was grossed out.

Certain Buddhists still have placards outside temples that bleeding women shouldn't enter. I recently saw such a sign outside a stunning temple in the heart of modern buzzing Hong Kong, of all places.

In Uganda, particular tribes still ban menstruating women from drinking cows' milk because they could contaminate the entire herd. And in Nepal, right now, menstruating women and girls are relegated to thatched sheds outside the home and are prevented from visiting others, in a charming practice known as 'Chapadi', because it's believed that a bleeding woman in contact with people or animals will cause illness and is just wrong.

Most Hindu temples also ban women from entering when they are bleeding. Some go further by banning women of menstruating age altogether. A particularly eye-opening case made headlines in India in 2018, when activists were successful in getting the country's Supreme Court to overrule such a ban at one of Hinduism's holiest temples. Historically, the Sabarimala temple has not allowed women between the ages of ten and fifty to attend because they could be menstruating and therefore will be unclean. Violent protests broke out at the temple after the ban was lifted, with many extremely angry men accusing the courts and politicians of trying to 'destroy their culture and religion' by allowing menstruating women access to what should be a peaceful place of prayer. Crowds tried to block any plucky female worshippers, female journalists were attacked and one of the women who tried to attend the temple ended up needing a police escort.

The saddest part of the whole unnecessary debacle?

The number of women who attended the protest in support of their own ban. That's how deep these myths can run. Women can be so convinced by men of their own filth that they turn on other women.

Devastatingly, Hindu girls and women also miss out on mourning their loved ones while menstruating because of this type of temple ban. Instead, they have to stay at home while the rest of the family pay their respects. Or, in the case of BBC journalist Megha Mohan, loiter outside the temple in Rameswaram (an island off the south Indian state of Tamil Nadu) as her family observed the final ritual for her grandmother.

While waiting, she recalls in a piece for the BBC News website, texting a female cousin, who couldn't make it to the final ritual, to tell her about her aunt stopping her from going to temple as she had asked for a sanitary towel:

> She sympathised with me and then she paused, typing for a few moments. 'You shouldn't have told them you were on your period,' she wrote, finally. 'They wouldn't have known.'
>
> 'Have you been to the temple on your period?' I asked.
>
> 'Most women our age have,' she said casually and, contradicting my aunt's earlier statement just half an hour earlier. She added, 'It's not that big a deal if no one knows.'

So, she could have just lied. And probably should have done to get her way.

Looking back further, Roman philosopher Pliny the Elder wrote in AD 60 that having sex with a woman on her period during a solar eclipse could prove deadly.

Sure, mate. But this example shows just how long we have lived with this ritualised shaming of women.

Jump forward two millennia, back to London and my wedding lessons – it was so very odd to hear such backward advice still underpinning the way women are made to feel today. In the developed world, religion may not have the control it once had, but it's still a huge cultural force that shapes norms and makes people feel a certain way about things.

Often, as with so many other day to day battles that make us weary, we can zone out when a religious teacher says something we would normally question. But keeping this shaming, impure narrative around periods going within our oldest religious institutions, despite leaps in medical advancements, at the very least has the subliminal effect of making women feel dirty and wrong.

And it reinforces the idea amongst men that our periods are something to fear and be sickened by.

CALL IT OUT, WARRIOR

So, question the nonsense when you happen to hear it. By all means laugh at it. Make sure you ridicule anyone who tells you that you are impure for your body doing something so natural that the entire human race depends on it. It isn't easy. I failed to do so out of a mixture of shock and a fear of offending the kind people trying to prepare me for my wedding. And I still regret it.

You are a warrior – who bleeds and goes to work every day. Not some dirty hermit who deserves to be quarantined and needs male approval to be reintegrated into mainstream society after your monthly bleed.

Religion has no borders. It was viral before the internet. That is why a privileged educated woman like myself can be told, while living in one of the most advanced societies on earth, not to hand my husband a piece a steak I've made for him while menstruating. I am about as far away as you could get from a girl in Nepal banished to menstrual huts away from her home while she bleeds. And yet, we ended up getting a similar memo. Except I have the tools, power and voice to push back.

You see, without us realising it, these myths permeate, settle and erode confidences. Keep your antennae up and tuned. And have the confidence to belly laugh and challenge the shaming beliefs in religion.

Our periods hold the key to bringing the next generation of society into being. The least we can do is make sure we have the right attitudes towards them and diagnose any lingering bullshit from the days when only men had the power to tell the stories that narrated and controlled our destinies, regardless of whether they understood us or not.

Today, women control our bodies and our narrative.

We must not lose control of that hard-won right – especially over our periods – at a time when our voices are louder than ever before.

You can choose your reaction. That's a power which must not be forgotten. Don't internalise any of the shame these myths propagate. And remember to actively call out nonsense when you hear it.

CHAPTER THREE

BAD BLOOD

'What if I forget to flush the toilet and there's a tampon in there? And not like a cute, oh, it's a tampon, it's the last day. I'm talking like a crime scene tampon. Like Red Wedding, *Game of Thrones*, like a Quentin Tarantino *Django*, like, a real motherfucker of a tampon.'

Amy Schumer, Trainwreck

It's time to focus on the group of people who are nearly as good as the men at period shaming.

Women.

It definitely isn't our fault the way society is set up to be only horrified or titillated by women's bodies. Nor is it our fault that we are the ones tasked with physically producing the next generation (the very reason for periods in the first place) – a role which tests our bodies and minds in all sorts of unacknowledged and undervalued ways.

But when you live and breathe in the bubble which normalises such attitudes, you internalise them and make them your own. Which means women end up feeling ashamed of a perfectly natural bodily process, often ignoring their bodies' cries for help and, in turn, shaming other women too.

Remember the confession booth we built for my radio show? I'll never forget the softly spoken woman in her twenties who poured this truth into my ear:

'Periods suck. We women are complicit in the silence.'

She isn't wrong. We are complicit. And such desire to stay silent about our monthly bleeds leads to all sorts of ludicrous scenarios and some very serious ones too, which I will come onto with my own near-miss situation.

But let's start with an absurd tale, one which perfectly sums up how women can be their own worst enemies when it comes to making periods taboo.

WOULD YOU GO TO PRISON FOR YOUR PERIOD?

I only inquire because one woman nearly did time in the can, simply because she couldn't bear to confess she was menstruating.

Let me tell you about the Canadian performer, Jillian Welsh. She poured her heart out to producer Diane Wu on the hugely popular podcast *This American Life* about a bloody evening scorched onto her brain and has kindly given me permission to reproduce her story in this book, aptly signing off her note to me 'yours in blood' (I love her already). The episode was focused on romance and how rom com scripts would play out in real life. Or not, as the case may be.

Jillian was twenty and studying theatre in New York when she met and fell for Jeffrey, whom she was starring alongside in a Shakespeare production. Fast forward to the wrap party and the cast night out. One thing led to another, they kissed and ended up back at his place. So far, so good.

Except Jillian's Aunt Flo was in town. Due to her highly conservative background she couldn't bring herself to even say the word period, let alone tell her new beau that she couldn't do the dirty because she was menstruating. But, finally, she fessed up – and guess what? He didn't care. Excellent sexy time ensued, after which Jeffrey went for his postcoital wee and shower, flicking the light on as he exited bedroom stage left.

As Jillian recounted to *This American Life*:

> It looks like a crime scene. There is blood everywhere. This is the first time I had seen so much of my own menstrual fluid. I was afraid of it. I couldn't even fathom what he was going to think about it... And then I don't know how this happened, but my very own red, bloody hand print is on his white wall... He didn't have any water or anything in his room, so I used my own saliva to wipe the bloody hand print off of the wall, like, out, out, damn spot.

OK let's pause there. It's grim but not that grim. However, it gets worse. Deliciously so.

Jillian then decided the best strategy to deal with Jeffrey's desecrated bedsheets was to stuff them into her rucksack, because she couldn't bear the idea of him having to wash them. She then covered his bed with his throw and prepared to scarper as soon as he was back from his shower. She offered a lame excuse, he looked suitably hurt and off she trotted to the subway, upset and laden with stained, stolen sheets.

> Then it really hits me that I have stolen this man's sheets. How do you come back from that? How do you – how are you not the weird girl who took his bedsheets? . . . So then I'm so inside myself and I hear this voice being like, 'Ma'am, excuse me, ma'am.' And I look up. And in New York, they have this station outside of subway entrances with this folding table and the NYPD stands behind. And it's a random bag search.

Let's pause again. *What would you do?* I know for certain I'd brick myself as soon as I was aware I looked like a murderer on the underground.

Jillian also panicked and pretended not to hear the officers, playing that, 'I am invisible game' you enact as a kid when there is nowhere left to run and you just hope by praying hard enough no one can see you anymore. She left the subway with a quickening pace. But to no avail. The officer soon caught up with an increasingly suspicious looking Jillian, opened her rucksack and saw the fruits of her sexual labour: crusty blood-soaked sheets.

> I remember him – and the subway has such distinct lighting – like I just remember him holding up these sheets, my menstrual sheets of shame, like menstrual sheets of doom. I realise that they didn't look like menstrual sheets of doom, they looked like murder sheets of doom.He asked me to explain it, and I just start crying. And I can barely get the words out. I'm just trying to explain to him, it's my period on those sheets. And I stole the sheets from the guy that I was with. And I know that that's wrong.

Now, when I asked you if you would go to prison for your period, you might have laughed, but Jillian's shame nearly led her down that road. Because, these two cops offered her an ultimatum: either go with them to the local police station, where they would file a report and ask her more questions, or take them (and the bedsheets) back to hot Jeffrey's house to corroborate her story.

It is what Jillian confesses to *This American Life* next which I find so fascinating:

> And I had to think about it . . . I honestly gave it a really solid, good think. There was a huge part of me that would rather go to the police station than have to go back and show Jeffrey these – not only show him these sheets, but also bring the police there. But, you know, my common

> sense caught up with me because this looks like I've done something very wrong.

Fortunately, Jeffrey, like the sexy period hero he is, when confronted by the cops, a nervous Jillian and the bloodied bedsheets on his doorstep, verified her story. Without skipping a beat, he simply explained that the sheets were covered with 'menstrual fluid'. No shame. No juvenile euphemism.

Jillian, as you would expect, is by now a sobbing mess and in a line which could have come straight out of a Richard Curtis movie script, he calls her 'wonderfully strange'.

Spoiler alert: if you're interested in finding out whether their love affair worked out, it didn't. Period night didn't kill the relationship, it was actually American visa issues. But it's not their love story that we're focused on here, what I care about is that a woman – in one of the best first sex stories I've ever heard – was so ashamed of her period that she nearly chose a night in the police station over returning to the 'scene of the crime'.

Take that in. It's bonkers. Fully bonkers. But you know what's even more crazy? Women the world over will understand why the police station inquisition was a serious option for a fully innocent Jillian because it seems we all have the propensity to become liars and weird little thieves when we get our periods. Anything to simply hide the evidence.

Take another woman I know, who also robbed some bedsheets. Jane was in her final year at school when she came on her period during a night out and didn't have any tampons with her. She deployed ye olde faithful technique of stuffing one's knickers with tissues and hoped for the best. Crashing at male mate's family house for the evening, she woke up the following morning to her own crime scene spread across the bedsheets. Just because her friend was a guy, she felt she couldn't talk to him about it. So, just like Jillian, she robbed the sheet, stuffed it into her handbag and then chucked it into a public bin on the way home. To this day, her mate's mum still asks for her sheet back, and Jane is too embarrassed to tell her the truth.

Linen is never safe around a menstruating woman, but particularly, it seems, around a woman who is ashamed of her own blood.

We also become super sleuth laundry women. Another woman I know, now an accomplished doctor in America, had to steal and sneakily return a guy's jeans so she could wash them:

> My worst period story was probably in college, I had my period and needed to change my tampon but hadn't yet – my then boyfriend came in to my dorm room and pulled me onto his lap… I'm sure you can see where this is going. I'm pretty sure I realised that I was sort of leaking through and then decided I just had to stay there forever. But eventually (obviously) I stood up and there was a real life *Superbad* moment AND I WANTED TO DIE. But actually, I just stole his jeans and immediately

washed them, returned them and said nothing about it. Truly horrifying.

You get the picture. Ludicrous behaviour abounds in women from all backgrounds and of all ages. All over some spilt blood.

And yet there is a serious level of irony that most young girls crave their first period, fretting about when they can join the 'P-Club' but spend the rest of their lives covering it up.

For one of my friends, this happened almost immediately. She'd just turned thirteen when her first period started, and her initial reaction was 'BEHOLD ME, NOW I AM ALL WOMAN'. However, this was somewhat tempered by the fact that she was on a five-day school trip to the countryside and had to figure out how to climb down a rope frame without anyone realising she was bleeding (whilst simultaneously giving off the laid back, mature vibe of one who has just 'become a woman').

Crucially, I raise this mad urge towards concealment not because I think women should be talking about their periods all time, but because this culture can harm women's health when they fail to seek diagnosis for menstrual conditions or gynaecological problems and furthers the stigma around periods – so it's time to shine a glaring spotlight on this silence and our bloodied sheets.

SO, DO WE *NEED* TO HAVE PERIODS?

Let's take a step back for a minute and consider: what is the point of a period?

Other than the important business of reproducing, according to most doctors there is very little point. Galling, isn't it?

Considering that this bleeding window in our lives is a relatively short amount of time – and only for those women who *want* or *try* to have babies – we are spending a heck of a lot of time and effort bleeding, when perhaps we don't have to. (Not to mention the energy expended hiding this natural process from colleagues, friends and other halves.)

Dr Jane Dickson, the straight-talking vice president of the UK's Royal College of Obstetricians and Gynaecologists' Faculty of Sexual and Reproductive Healthcare tells me:

> A woman is built around her reproductive cycles. She is set up as a pregnancy machine . . . A period is a natural, in-built preparation system for pregnancy. But in this day and age there is no reason a woman should have periods if they don't want them. It's totally healthy to use contraceptives which stop bleeds altogether or create artificial periods.

Moreover, (and I hate to break it to you) artificial periods, the ones you have on many pills during the seven-day break, are also hangovers from an even more puritan age.

> There is no reason for a one week break [within which to bleed] any more either. When the pill was first developed, it contained an extremely high dose of hormones – five times what the modern day pill contains now. It made many women feel sick and unwell. So, they liked the idea of a seven-day break from the heavy hormones.
>
> But the pill was also developed in America – a heavily Catholic society – where contraception was frowned upon. If women could still have periods while on the pill, they could mask the fact they were using a contraceptive and it would be less stigmatising. And women themselves were reassured by seeing a period every month as healthy menstrual function.
>
> As science has developed and the dosage is now greatly reduced – and contraception in many parts of the world is far less stigmatised – none of those reasons for a bleed exist any more. The pill just switches your ovaries off and keeps the womb lining suppressed. The injection dupes the body into thinking it's pregnant; the Mirena coil suppresses the period – there is no point having a period whatsoever other than when you want to reproduce.

In fact, while writing this book, the official health guidance in the UK changed, finally revealing to women on the pill that they no longer needed to take the traditional break to have a bleed. I quote the guidance: 'There is no health benefit from the seven-day hormone-free interval.' And, 'women can safely take fewer (or no) hormone-free intervals to avoid monthly bleeds, cramps and other symptoms.'

This is game-changing. And very overdue.

Pill-taking women across the world erupted in shocked and righteous anger at the news. For decades, women had been bleeding when they didn't *need* to. It's ludicrous. Why has it taken until 2019 for the official health advice to tell them their pill-periods were non-sense?

And the real red rag to the raging bull? Those fake bleeds were designed to make an old man in a white hat happy. Yup. It all comes back to the Pope. Professor John Guillebaud, a professor of reproductive health at University College London, told the *Sunday Telegraph* that gynaecologist John Rock suggested the break in the 1950s 'because he hoped that the Pope would accept the pill and make it acceptable for Catholics to use'. Rock thought if it did imitate the natural cycle then the Pope would accept it.'

For more than six decades most women have unknowingly been taking the pill in a way that inconveniences them in order to keep the Pope happy. Digest that. Sticks in the throat a little eh?

Women felt and feel rightly duped. Another sodding period lie told for more than half a century to benefit someone other than the bleeding woman suffering unnecessarily.

Of course, some women don't want hormones in their body. They like being natural. They *want* to bleed – regardless of ovarian intention. Some argue the time of the month is a source of strength for them, or perhaps they can't find a pill or contraceptive which doesn't make them feel ropey. Others argue that ovulating naturally is good for one's health, as is the natural production of the progesterone and oestrogen.

I ventured to Dr Dickson that perhaps a period is useful as a marker of health or ill health – and she batted away my concern with that easy breeziness and reassuring aura of a fact-laden specialist. She explained to me that there are usually other symptoms to other illnesses which don't require periods as a signifier. I.e., if you had cancer of the womb, you would bleed anyway even if you were on a pill that stopped you bleeding; or polycystic ovaries would manifest through a range of other factors, such as excess hair growth or loss due to overactive male hormones.

MY OWN PERIOD CONFESSION

It's at this point I have something to confess to you.

While penning a book about periods I haven't had a single one. Not so much as a menstrual splash until the very last chapter (you're in for a treat). It feels odd, dishonest somehow, though I've definitely done my time in the menstrual trenches. I have indeed bled while writing.

I did that for about six weeks. But it wasn't menstrual. My period hiatus is because I've been pregnant. Pregnant with a baby I could have so easily missed out on having. And now, despite all of my doom-infused expectations I've had said baby (whoop!) – hence the six-week post-birth bleed. And then, because I've been breastfeeding the beauteous wonder that is our son, while battling mastitis (the vile blocked duct breastfeeding infection), I have yet to bleed naturally.

I have already told you it took two decades for me to be diagnosed with endometriosis, a debilitating period disorder that affects one in ten women – including Marilyn Monroe, Hilary Mantel and Lena Dunham, if you want to know the A-List.

I have already told you of my bewilderment and shame that I, a vociferous woman who loves asking tough questions and soliciting the truest answers I can, had failed again and again to secure a diagnosis – despite having traipsed in and out of doctors and gynaecologists over the years complaining of severe pain.

But what I haven't told you is how my lack of diagnosis could have cost me and my husband the chance to have a biological baby.

And my experience strikes at the heart of why I am urging women to drop the shame a male-organised society has foisted upon us: our wellbeing and health.

But make no mistake, I am not preaching about the need to drop the period stigma so women are more clued up about their fertility. (Although this is a desirable side effect should women care to have children.)

This is not some kind of dystopian *Handmaid's Tale* plot twist.

Instead, I want to share what happened to me as a way of highlighting how knowledge is power. And how important it is to be as unashamed as possible, so we keep pushing for answers about areas of women's health which have for centuries existed in the shadows. Too many women are simply soldiering on while struggling with all sorts of gynaecological and sexual issues because they think that is their lot in life.

Bluntly put, often we put up with our internal lady piping and vaginas not quite working as they should because we are embarrassed and we don't believe it to be our absolute right for everything to be more than *all right.*

Nor are we always believed by the doctors who listen to our woes – once we muster up the energy to try to communicate our problems.

I always knew something was wrong with me gynaecologically but I too soldiered on. From age eleven, these heavy painful periods were the norm. I tried everything – strong painkillers, drinking coffee (which I loathed but my mum had heard could help), furry hot water bottles, lying on my stomach, and then, as I grew older, getting just that little bit more drunk on nights out when I was either about to start or in full flow. And then at university, I finally found a pill which agreed with me (although not the large quantities of booze I was imbibing – I went from being an excellent loving drunk to quite the vile bitch drama queen). Hence began the great cover-up as I like to call it.

From the age of twenty-one to thirty, I happily chomped my way through pill packet after pill packet, and my periods, although still uncomfortable, became more manageable. But when my husband and I decided perhaps we should start thinking about having a baby and I ditched the pill, my real periods – the dark bastards – really started to return.

We couldn't get pregnant. We seemed to have gone from a place of not wanting a baby urgently, to everyone around me falling pregnant. Suddenly I was in a place I'd always feared: infertility. Because somewhere deep down, I knew I wasn't quite right, even though every doctor and specialist told me I merely had a bad case of dysmenorrhoea, a fancy word for painful periods. With a great sense of foreboding, I had secretly dreaded the stage of my life when I would attempt to produce life, convinced something might be wrong. And here I was. And it wasn't going well. Far from it.

As each month went by during our two years of trying, my periods were getting worse. They were starting to reveal themselves in their full natural horror, free of the contraceptive mask which had been restraining them for the last decade. The lowest point I can remember on this knackering journey was during a holiday in Sweden, from whence my mother-in-law hails, walking behind her, my father-in-law and my husband, after a sunny coffee and cinnamon bun pit stop on a picnic bench. I felt like iron chains were dragging my stomach down, pulling me towards the floor, as my bones ground against each other during what should have been a lovely easy amble around a Stockholm park. I came to a complete stop, unable to take another step. I just couldn't move anymore. The period pain was so great. I stumbled to the closest bench and didn't move for a long, long time.

That was the moment I knew I needed some cold hard medicine. Not some muddy herbal tea nonsense from an overpriced acupuncturist. Nor another expensive and pointless colonic irrigation that did nothing other than to make me feel lightheaded and immediately crave a greasy cheese and caramelised onion toastie.

On that day, I admitted the first of two defeats. The first was that something was wrong with me to the extent that action was required. A week later I was booked in for a laparoscopy, a keyhole procedure which serves both as a diagnostic tool and a treatment for endometriosis. So, you sign off having a diagnosis *and* treatment while you are under anaesthetic, meaning you either wake up after a few minutes as no action was required or after a few hours as the doctor has been beavering away.

I was the latter. I did indeed have it, endometrium (old womb lining which should leave one's body during a period), coating my organs, mainly my bowel and bladder, but very luckily it hadn't stuck to my ovaries, uterus or fallopian tubes. The disease was at stage two of four. Moreover, after two and half hours of painful lasering (during which they inflate your organs with air) the doctor felt he had managed to remove all of it.

In the six months after a laparoscopy, women who have struggled to conceive naturally because of endo have a much higher chance of doing so. And the debilitating pain can go away or be significantly reduced. Sadly, despite our best efforts – and they really were Herculean – pregnancy still wasn't happening and my periods were as punishing as ever.

Hence came my second defeat, as I stupidly and naively chose to view it at the time: I agreed to IVF. Not being able to fall pregnant felt like a huge failure. We'd been told repeatedly that our infertility was unexplained, so I had strongly resisted the idea of IVF previously because having such a major intervention felt like I really *had* failed and that we'd reached the end of the road.

An amazing older female doctor in the NHS promptly disabused me of this foolish opinion. She met with me and my husband for an appointment regarding our ongoing issues and my recovery following the removal of my endometriosis. All I really remember about this meeting with this wise, stern but kind doc, was her shiny grey ponytail and her saying something along the lines of: 'For God's sake Emma, stop being so stubborn and just have IVF. You've qualified for it on the NHS for a long time now, especially because you have endometriosis and have tried for more than two years to get pregnant. What have you got to lose? Your periods are awful each month and this is one way of trying to stop them and get pregnant before. It's really a win-win situation.'

Before I knew it, despite all of my reservations about the hormones, the intervention, the hope, the potential crushing disappointment and the overarching feeling of total failure, I found myself saying yes.

The doctor sold the process to me on potentially ending my periods for a little while (that's how bad they were) and I consented to the hormonal rollercoaster that is IVF without a millisecond of further contemplation. I didn't for one moment dare to think it might work and produce a baby. No, that would be far too easy and require a dose of luck I didn't seem to possess with my wrecked body.

I remember my husband saying: 'Emma, wait. Shouldn't we discuss this?' As we were both bundled off for various blood samples, clutching consent papers. I turned back to him and simply replied: 'No. I can't go on like this.' He went with it unquestioningly, because he's a legend.

We had a big holiday already in the diary: three weeks travelling around China (another 'we can't get pregnant' adventure holiday – where you splurge money on an amazing voyage in a bid to distract yourselves from the pain of unexplained infertility). Hero doctor agreed to wait until we came home. By this point, I was totally in her hands and meekly agreed to everything she said. It just felt so good to have someone else take control and bring some order to my menstrual and mental chaos, caused by not being able to conceive and not being able to exist easily in my pain-wracked body for one week every month.

Another potential fly in the ointment was to come, though. During our last few, carefree days in China, the 2017 snap general election was called. I knew a punishing but exciting work schedule greeted me when I returned home, as I would have to go on the road with work chatting to voters and interviewing politicians up and down the country. I thought with a mixture of relief and resignation, IVF would have to wait.

Hero doctor was having none of it when we spoke and I started to make my excuses. 'We can make it all very portable for you and explain how to do each stage on the move. You can continue as normal.'

That was that then.

Fast forward two weeks, and I'd secretly loaded a mini fridge in Skegness with my secret injections, as I tried to hunt down Paul Nuttall, the then-UKIP leader, on the election trail before presenting my three-hour show by the seaside in glaring hot sun. (That's another thing women don't talk a lot about: IVF. Well, certainly not while they are going through it. I was so convinced it wouldn't work that I couldn't bear to mention it to my nearest and dearest friends and family, never mind my work colleagues, as I didn't want to make it real and then have to deal with more crushing disappointment when it didn't work.) The injections went ahead successfully, unlike my interview with Mr Nuttall, who was suddenly nowhere to be found and had stopped taking my calls. A familiar story.

On one of the final days of my forty day injection marathon, and nine days before the results of the general election, which Jeremy Corbyn hoped would make him Prime Minister, I got the chance to speak to him about his vision for the country on Radio 4's *Woman's Hour*. Much has been written about that interview, as Mr Corbyn had expressly come on the programme to announce his party's flagship childcare policy, only to not know how much it would cost when repeatedly questioned, despite being armed with his manifesto booklet and his iPad, both of which he consulted several times. Our exchange duly went viral, the reactions poured in (including some extraordinary anti-Semitic abuse which Corbyn condemned) and I somehow stayed calm and kept injecting my way around the country and political candidates.

Just over a week later, this time of injecting culmniated in my exclusive sit down interview with an ashen-faced Theresa May in her office at Downing Street, who admitted she'd cried on the night of the result as it dawned on her that the biggest political gamble of her life had spectacularly backfired. She lost her majority in the House of Commons. I listened to the Prime Minister intently, pushing her for answers about her leadership and strategy – all the while fully loaded with hormones.

The reality was the 2017 general election was the

very best distraction I could have wished for. All this claptrap about taking it easy when trying to conceive hadn't worked for me in the previous two years, so I decided not to try to stop my life while filling my system with egg-inducing hormones. I ploughed on, kept my vow of silence on the matter and only had one day of feeling distinctly odd, which luckily was a Saturday where I wasn't working.

The only terrifyingly exciting moment came when peeing on a stick at 5a.m. two weeks after the implantation of the embryo that was to become our darling boy. I had been fine up until that point because I hadn't let myself believe. I am a realist who deals in fact. And I was in full self-preservation mode. So much so, that when the test showed positive, I still didn't believe it. That was until I sent a photo of my pregnancy stick to my doctor friend (the same friend who had diagnosed me with endo in the first place), who confirmed we had indeed managed to fall pregnant.

Against all my self-imposed negative odds, I had a child inside me. I was quietly and very nervously thrilled. As was my husband. Our luck had finally come after years of trying – and we knew all too well just how fortunate we had been for IVF to work at all, and first time at that.

I had left the grim period camp behind. At least for now.

To my surprise, a relatively straightforward pregnancy ensued (again more amazing luck – gratefully received and regularly acknowledged), with the only noteworthy blight a night in hospital after some bleeding due to my temporarily low-lying placenta. Despite my growing wriggly bump, I still didn't really let myself believe it was happening until I was lying on the operating table nine months later, having my son pushed and pulled out of me during a C-section. The miracle is still dawning on me, as I blearily get to grips with new parenthood and get to know our wide-eyed, strong-willed giggling son (whose first word may just be period, having been stuck to me while writing this book).

Let me say this: I had always wanted children. That was never in doubt for me or my husband. But we only started rather idly trying for a child because of a gap in my work schedule. An opportune moment had suddenly presented itself. At that point we weren't desperate for a child in our lives, but as nothing happened and my periods worsened and worsened after coming off the pill, our rather out of the blue decision to go for it ended up leading us down a rabbit's hole of pain, frustration, infertility and on my part, worsening health – with no explanation – until diagnosis.

If we hadn't randomly decided to start trying for a child at a slightly earlier stage than either of us had anticipated, my endometriosis could have had more time to develop and stick itself to more organs, including my ovaries. I had taken to suffering in silence, as so many women do, especially when GPs and specialists alike have told you all is well and paracetamol is your best friend.

My point is, it was pure luck that I started trying for a baby earlier than I had thought we would; that my friend is an obstetrician and had seen me during that time of the month, and that my endometriosis hadn't spread to my fertility organs.

Knowledge is power, but how can we access that knowledge if we are suffering in silence? Women can be their own worst enemies. Me included.

My soldiering on for just a few more years could have cost me and my other half the chance to have a biological baby, and I didn't even know what I was soldiering on into. Staying silent about my difficult periods after being disbelieved and fobbed off with painkillers for years and years would have had major consequences for me and my other half. We were sleepwalking into potential full infertility without either of us knowing it. And that's just one cost of women not fighting repeatedly for answers when their gynaecological health isn't quite right. And of doctors not believing them.

But I am not alone. Far from it.

WOMEN AREN'T HELPING THEMSELVES MEDICALLY

A major report by Public Health England out in 2018 highlighted for the first time just how many women suffered some kind of sexual dysfunction and reproductive issues, but are not seeking help. 42 per cent of the thousands of women interviewed said they do not enjoy sex – and that number is highest amongst the young. Digest that. Young women aren't enjoying sex and are doing nothing about it.

Moreover, a significant 31 per cent had experienced severe health symptoms in the previous year, including heavy menstrual bleeding, menopause, incontinence and infertility. The authors of the report stressed the 'impact these issues have on women's ability to work and go about their daily lives', something that's never been shown in such a report before. Finally, women are being asked these questions, and it's painting a bleak picture.

Sue Mann, a public health consultant in reproductive health at Public Health England, said there was still a 'stigma associated with talking about reproductive health issues' but wanted to make the point, it doesn't need to be like this. 'Enjoying a fulfilling sex life is important for women's mental and emotional wellbeing. Our data shows that sexual enjoyment is a key part of good reproductive health.'

Dr Dickson (whom we met earlier in this chapter – the straight-talking vice president of the UK's Royal College of Obstetricians and Gynaecologists' Faculty of Sexual and Reproductive Healthcare), was also involved with the report. She added:

> The importance of having a healthy, enjoyable sexual life cannot be overstated as this strongly contributes to general wellbeing. However, there is still much stigma and embarrassment when it comes to sexual

function – especially when we are talking about women's sexual pleasure.

Society still relegates women's sexual pleasure to the background in comparison to the importance assigned to the gendered roles that women carry – such as that of being mothers. If women lack sexual enjoyment they should know that they can talk to a specialist and get support from psychosexual services in sexual and reproductive healthcare clinics.

Women battling on through shame, lack of time and being disbelieved needs to stop. Just as it's our right to vote, have clean water and share our opinions, women need to start believing in their right to a decent quality of life when it comes to our periods, sex lives and general vaginal health.

Dr Dickson believes that, as a society, we think we have to be actually sick to see the doctor. Anything which threatens our wellbeing isn't considered serious enough.

Women are more likely to see the doctor about aching knees or migraines because they are deemed proper illnesses and problems. But we don't focus enough on wellbeing, as opposed to illness and if something isn't right with periods or our sexual health, women don't want to waste the doctor's time. And yet if we focused more on wellbeing, we would have less sickness.

There are many cases where women present with gynaecological problems unacceptably late because of shame and they didn't think it worth bothering the doctor.

It is true that if we prioritised feeling well in our vagina and its connected organs, issues may not worsen into full blown sickness. This reticence about our female parts *must* end – not simply because women deserve a decent quality of life but because greater pressure from greater numbers will force more money into research around these issues.

For instance, endometriosis is as common as type 2 diabetes, but research into it is scant and still nobody knows what causes it and how to control it. For every person with type 2 diabetes in the US, the American National Institute of Health spends $35 a year in research funding; for every woman battling endo? Less than $1.

It may seem like a cheap joke, but if men had periods there would probably be a cure by now. The majority of scientists, until relatively recently, have been male. It makes horrible sense that men have prioritised issues *they* understand or the whole population could be threatened by, but now women need to speak up en masse – they need to share their symptoms and help exert pressure so the few researchers and charities like the superstars at UK's Wellbeing of Women, who are attempting to understand uniquely female health issues, have an army on their side.

Oh, and when women do find the time and energy to schlep to the GPs? They need to be taken seriously and believed by the overstretched doctors on the frontline. This is key. And they need to keep going and seeing different doctors if problems persist.

THE CRIMSON TIDE IS TURNING

Things *are* changing. A few women, here and there, are eschewing shame and daring to show the world what's been going on privately in their pants for decades. The fact that every time they do it's headline news is telling in itself.

Remember Kiran Gandhi? The 'freebleeding' marathon runner labelled 'brave' and a 'menstrual hero' for daring to let her period flow down her leggings as she pelted it around London in 2015? If nothing else, you will probably be able to recall the image of her with a serious trickle of blood in her crotch. I told you that we'd meet her again properly.

Kiran had got her period the morning of the run – a run she had been in training for with two friends for months, just like thousands others who had made the trip to the famous start line. The Harvard graduate, activist and musician had a choice: abandon her run (as she had stopped her training whenever her period came) or just go for it, but this time without the inhibiting chafe and discomfort a sanitary pad or tampon would cause.

'I remember thinking I would rather bleed freely,' she told me. 'I knew this was a radical choice as free bleeding wasn't something I just normally did.'

No, none of us do, unless we have no idea our period is there in the first place. Or, as we'll discuss later in the book, we have no choice because we don't have the financial means to buy any protection.

'I just went for it and it felt completely normal. I was already sweaty so I didn't really notice the blood that much. And even though it was day one, which is usually the heaviest and most uncomfortable, I managed it.'

I recall seeing photos of Kiran at the finish line with a triangle of blood down her red running leggings and in the same thought thinking 'Wow', and also 'Oh yeah, of course that's what your period would look like on your clothes if you just didn't wear anything'. But while strangers around the world reacted in their droves to her decision (many positive but also many negative, calling her disgusting and unhygienic), Kiran crossed the finishing line to be met by her dad and brother, who had definitely not seen her period blood before.

'My theory was you can't shame a marathon runner, you are doing something so amazing with your body. And bleeding. But I definitely had internalised some shame about my period. And my brother and dad had never seen me like that before,' she confided to me. 'I remember my dad asking if I was OK. It was good for him to see my blood as all we women do is hide it. It was a really good thing for him and my brother to see what women can do – even while bleeding. My dad switched any shock or disgust he may have had to being more supportive and inspired.'

Bluntly put, Kiran shouldn't be seen as a hero. And

yet it speaks volumes about our society that it was such a global story simply for a woman to let her body do what it naturally does without the inhibiting aid of absorbent cotton wedges. I am reliably told marathons already include a lot of chafing around the inner thigh and nipple area – who the hell wants some additional vagina chafe too?

Similarly, when the poet and artist Rupi Kaur decided to post a menstruation-themed photo onto Instagram in 2015, a similar moment of open mouthed horror collectively ensued.

The photo in question, taken by her sister, showed Rupi, fully dressed, lying on her side, facing away from the camera, with a small blood stain on her grey joggers and on the bed next to her. The photo isn't remarkable in any way to millions of women the world over, as it conveys a sight and feeling so familiar to us in private (as we reach for some Fairy Liquid or detergent to quickly scrub out our annoying leaks from our pants or bedsheets). And yet it was enough for Instagram to censor it and take it down from the photo-sharing site. Twice. Because it didn't follow their community guideline. You know, the same guidelines that prohibit sexual acts, violence and nudity – and say nothing about periods. Rupi's response on her Facebook page was note perfect. Here is part of it:

> Their patriarchy is leaking.
> Their misogyny is leaking.
> We will not be censored.

> I bleed each month to help make humankind a possibility ... Whether I choose to create or not. But very few times it is seen that way... We menstruate and they see it as dirty... As if this process is less natural than breathing... As if this process is not love.

I am not preaching that you should suddenly go on a run in your local park and bleed all the way down your leg, then shake your stained lycra leggings in the face of your male relatives when you arrive home. Nor am I saying take a photo of your worst period stain and post it online in a bid to garner some likes and unhappy emojis. Although, if you want to, *do it*! Society has always had activists and artists to take stances about things which should be normal and draw attention to them.

But you should think about why you don't see periods anywhere and why, when you do, they are still so shocking.

An older friend told me this story which I love because it is the antithesis of shame. In 1980 – long before social media allowed folk to post shocking, non-shocking period photos – she was at university, and was invited to a party with a loose fancy dress theme of 'dress for freedom, school's out for the summer'. Most people rocked up in short skirts, punk-esque make-up, ripped clothing and a clutch of washable tattoos. You can imagine the scene. But one woman, who was my friend's most respectable, quiet, well-behaved, serious and hard-working mate (who didn't usually even bother attending parties) turned up – and, in the most unassuming way one could manage to do this – wearing a used tampon on a ribbon around her neck.

Her understated but rather kick-ass rationale was that the tampon symbolised ultimate freedom for women in the eighties, but quite why said tampon had to be used was never explained. (It's interesting to pause and reflect on how Kiran Gandhi found a tampon inhibiting three decades later). My pal says the shock still resonates to this day. She's never forgotten it for two reasons: the sartorial decision seemed so massively out of character, but also because of the shock of a tampon in public view. With actual period blood on it. Remember, this was an era when girls were often told they couldn't wear a tampon until after they'd had sex, and if they did it was seen as a dirty admission that they were no longer a virgin. So maximum privacy was the order of the day around this part of life – especially if you were as 'loose' as to wear a tampon. Around your neck or up there.

Shame might not be as tangible as a dirty tampon necklace – but it pervades all the same. Stop fuelling the fire. Stop being complicit in the period silence.

You should think about that gynaecological pain you've had and whether it's normal. You should see a doctor if you hurt in your vagina when you have sex or your periods are unbearable. You aren't wasting their time. Like me, you could be losing precious time on a diagnosis that might mean the difference between having the option to have a child or not. Or simply having a better quality of life.

You should lose the shame about yourself and not contribute to any other women continuing to feel it too. Women need to stop being their own worst enemy on this front. Surround yourself by people who don't perpetuate such views and when a friend *does* speak out, support her.

You know if something isn't quite right in your body. You do. And I am not saying it's easy to get a diagnosis or help. It isn't. Take it from a woman who didn't get a much-needed diagnosis for 20 years and is embarrassed by that. But we can all stop being our own worst enemies and speak up about our periods whenever we need to – socially and medically.

And please – don't go to jail for your period. No matter how messy your period sex session was. OK?

CHAPTER FOUR

MAN BLOOD

'If men could menstruate, periods would be enviable.'
Gloria Steinem, American feminist, journalist and activist

It's time to confront the brooding 800-pound gorilla hovering over me as I write this book: men. And their period revulsion.

Newsflash – this isn't the male-bashing chapter, or even a big old moan about their often pathetic attitudes towards the very blood their lives depended on. Nor is it a worthy, handwringing cry for men to talk about something they have no interest in. I am not a fan of wasting my time or, more importantly, yours. Instead, I want to highlight why I think men have the issues they have with periods, to expose the true prejudice and recommend a workable solution.

The problem strikes to the very core of human nature. We are scared of what we cannot understand or experience – or have no interest in understanding. Hence, it makes total sense that men have 'othered' periods and, by extension, women. Men have made them weird, scary, mythical things that turn their stomachs and transform women into untrustworthy banshees.

Men don't get periods. They never have and they never will, but that doesn't mean that we women should allow their squeamishness to dictate the tone around our monthly bleeds any longer. Periods wouldn't be uncomfortable conversation fodder if there were only women stalking the planet – they would be the glorious norm.

It's time for women to take control of the chatter – or lack thereof – around our periods and take those unwilling men with us, whether they're kicking and screaming or just quiet, bewildered and in need of a biscuit and a sweet tea.

Imagine a walk you do every day. Perhaps it's from the train station to your office, or your house to the supermarket, or even a trundle to your favourite cafe. Now envision that same walk – past shops, fellow humans, cars, dogs – and consider leaving a trail of blood droplets behind you. Every step you take, a little or a large blood drop falls out of your vagina. Or maybe, if the dogs are in luck, a whole juicy clot crashes to the floor. (Sorry, too much? Stay with me, the struggle of a pre-Edwardian knickerless woman is real.)

Because, according to the excellent historian Greg Jenner, this is how women used to bleed before the sexy 'menstrual apron' was pioneered in Edwardian times. Greg writes on his website:

It's possibly quite shocking that not all our female ancestors seemed to have used pads, tampons, cups, or other devices to catch the blood. Indeed, many simply bled into their clothes, while others are said to have dripped droplets of blood as they walked, leaving a trail behind them.

Yes, you read that right. Long before knickers and sanitary towels were a thing, we women bled freely into our clothes, directly onto the streets or, if we were field workers, into the straw coming up around our knees. Take a moment to consider that. Of course great efforts were made to protect clothes and one's dignity with the fashioning of homemade rags or scraps of any vaguely absorbent material. Women would thriftily wash and rewash such rags over and over again as there were no commercially available solutions. But until relatively recently, sanitary ware was scrappy. Literally.

And even then we were trying to make it better for those non-bleeding folk around us, carrying nice smelling nosegays to help mask the rusty fetid smell emanating around our being as we free bled all over the shop. Literally. Free bleeding wasn't even the almost rebellious 'free bleeding' it's often seen as today – it was just *bleeding*. And it must have been noticeably pungent. And made a lot of washing for that mangle…

Of *course* men were bloody terrified.

Most of them probably had no idea what was going on! Except that women all around them were bleeding on the streets and no one could talk about it. And women, so busy trying to hide their blood, look after the children, cook the dinner, satisfy their husbands and potentially earn some money by sewing or some other suitable job, didn't have the *right* or the energy to wearily respond and calm everyone down. It's highly likely that they didn't even understand what was happening to their own bodies either – you think sex education is poor now? Take it back a century or so!

Eventually, in the 1800s, the menstrual apron came along for the more genteel lady (those who could afford such a thing) so as to avoid any more unseemly staining. Greg describes it as 'a washable linen nappy for the genitals, held in place by a girdle and joined at the rear by a protective rubber skirt. To ensure warmth and decency (if a sudden gust of wind lifted up her skirts) ankle-length knickers were also worn beneath the apparatus, but they would be special open-crotch pantalettes so no blood would stain them.'

The great leap forward came courtesy of busy but highly enterprising First World War nurses who used the wood pulp bandages applied to soldiers' wounds to absorb their own flow. Five times more absorbent than cotton, it was a key moment in the eventual invention of the sanitary pad. As if these women weren't busy enough stemming the flow of blood on others, they also found the wherewithal to come up with the sani towel prototype while tending to the needs of millions of terribly wounded and maimed men. Necessity truly is the mother of all invention for overworked and under-resourced nurses bleeding on the job and used to thinking on their feet.

I think we can all agree that Dr Earle Haas, aka 'Mr Tampon', did us all a major favour when he came along with his nifty product – the applicator tampon – in the 1930s.

The story goes that Dr Haas wished to create something that was significantly better than the rags his wife had to wear during her monthly bleeds. According to the Museum of Menstruation he was inspired by a friend in California 'who used a sponge in the vagina to absorb menstrual flow. So he developed a plug of cotton inserted by means of two cardboard tubes; he didn't want the woman to have to touch the cotton'. What a clever, clever man. He eventually sold the concept to another enterprising woman, Gertrude Schulte Tenderich, who patented it and began the company Tampax.

All women owe Earle and Gertrude a large debt. Yes their partnership ushered in the beginning of capitalising on our blood, a problematic phenomenon I will cover in more detail later on – but I am still bloody grateful someone did something to move women on from rags.

It's worth taking a moment to absorb (had to, sorry) the fact that these two products – the mighty tampon and the pad – have only been available on the mass market for just under one hundred years. That's right, even though women have had periods since the beginning of the human race, it's only in the last century that we've been able to soak them up effectively (most of the time). Even then, remember lots of women were culturally discouraged from using tampons for a long time – so full freedom to use tampons guilt-free is still a relatively new phenomenon. And yet many men are still horrified by the idea of periods – despite no longer seeing our blood droplets regularly and all of the secret aids we now possess in our period armoury.

Plus it's hardly as if we've redressed hundreds of years' fear with some overdue conversation that calms men's fears down. Thanks to better sanitary products, our periods are more hidden than ever – and the lack of leaks and centuries of silence mean that men continue to be both ignorant and appalled about what's going on down there. Frankly, they don't know how lucky they are not to be following a trail of blood down the meat aisle at Sainsbury's.

One of my closest friends recalls her twenty-five-year-old male housemate's horror, five years ago, when she left her tampon box out on the bathroom shelf above the loo. Heaven forfend.

Or, take my brilliant sixty-five-year-old father-in-law. He cannot even imagine how I am writing in any form on this 'awful' topic. Every time I bring up periods, he closes his eyes, furrows his brow and shudders, saying, 'Emma, please. Do you really have to?' To be fair, he is quite proper and squeamish about all bodily fluids – but he is also incredibly modern. Present at the birth of all three of his sons, he's always played an equal role in domestic duties, including happily changing nappies in a full suit and tie in the eighties. He just doesn't want to talk about what he deems 'unseemly matters' – and that definitely includes periods. Even though the major irony in this behaviour is staring right at him: the existence of his beloved children is (partially) courtesy of my wonderful mother-in-law's healthy menstrual cycle.

The fact is, many men of all ages and backgrounds still can't deal with acknowledging our menstruation. Of course, things must be slowly changing – with a whole generation of young boys growing up with the most emancipated mothers and sisters there have ever been. And yet, I wouldn't bet your bottom dollar on periods still coming fully out of the shadows, regardless of how kick-ass these women raising the next generation are. There's still a mafia-esque omerta which hangs around our cycle and the blood in our pants. Viscous, icky stuff that boys just don't want to know about and embarrassed teen girls, high on a cocktail of puberty hormones, don't feel they can talk about. And so, the cycle of silence just keeps on turning.

Or, the total opposite happens: instead of silence, deeply uncomfortable men start cracking out horrendous period jokes, turning slightly hysterical at the mere mention of menstruation.

Still not convinced that today's men aren't comfortable with periods – or are lacking period knowledge? Then you must check out these *genuine* questions that men aged between eighteen and forty put to journalist Hannah Ewens (and which she gamely answered in an article published on Vice) in 2017. The collective ignorance is astonishing, which is why I wanted you to see this list of enquiries. You have to know how bad a problem is before you can rectify it:

Do you like it when the tampon goes in?
Do you fuck during your period?
Are you most horny on your period?
Do women get angry on their period?
Have you ever been tempted to taste your blood?
Does it smell?
Does it suddenly stop?

(And my personal fave:)

Is it a lighter shade of red in summer – rosé – and a dark red – merlot – in winter?

Right then...

Men have never understood periods and seemingly still don't. They are a punchline or something to be feared or horrified by. And on the rarer occasions that they block men from potential orgasm, an irritation for the male species.

An old friend, Lee, generously shared his understanding of periods before he was married to a decent woman who refused to spare his blushes about such things: 'In school we learned from somewhere that periods were something to tease girls about. That they smell when they were "on". And it was a sign of frigidity too – as it was a reason a girl wouldn't do anything sexual with you.'

However, on this last point, girls couldn't win.

Because if we put out on our period, we become weird vampire sluts.

Sex during one's period is a theme we shall return to in a later chapter, but I want to highlight these recollections to show how certain men have become hysterical over periods. Oh, the irony. That very stick that's been used to beat women for centuries – that we lose our marbles in a hormonal rage at least once a month – is in fact a very appropriate description of the reaction of the opposite gender.

Such a reaction was perfectly illustrated by a forty-something male TV editor (who I love working with) when I told him what this book was about.

To my amazement, this enlightened fellow physically recoiled from me, turned a deep shade of period red and looked like he was about to retch. He then begged me to 'please change the subject as soon as possible'. However, despite his initial plea, ten minutes later, he still couldn't get off the topic – and I was struck by the fact that talking about periods seemed to send him temporarily insane. Suddenly he was cracking out his favourite period gags ('gags' is a touch generous) as

I sat bemused by his frankly unbelievable reaction, which bore no resemblance to his usual level-headed and intelligent behaviour. And we *women* are the supposedly hysterical ones on our period? From where I'm sitting, it's the men who go mad at the mere mention.

It's worth noting that the respected American psychologist, Leta Stetter Hollingworth, already settled this debate back in 1914. Her series of tests on both genders concluded that there was no empirical evidence to show women suffered any form of mental incapacity or decreased performance during menstruation.

It is men who have made women out to be dirty. For something perfectly natural. *They* are the ones who have turned periods into a filthy, terrifying freak show – as if it wasn't bad enough already! And they don't even have to clean anything up.

The current leader of the free world is one of the biggest public offenders. Opprobrium rained down on Donald J. Trump during his campaign to become the forty-fifth occupant of the White House after he basically accused the former Fox News anchor Megyn Kelly of being on her period when she gave him a firm but fair grilling during a Republican TV debate.

Kelly was grilling Trump over his misogynistic

comments about women – he had previously described some of us as 'fat pigs, dogs, slobs and disgusting animals'.

Candidate Trump wasn't a fan of the TV host's line of questioning, and in a post-show conversation with the CNN anchor Don Lemon, he said the following: 'You could see there was blood coming out of her eyes, blood coming out of her wherever.'

'Wherever' was quickly accepted as vagina. Of course, the business tycoon couldn't bring himself to mention such an unmentionable hellhole – or even utter the word pussy, his seemingly preferred moniker for female genitalia that he has boasted of 'grabbing'.

Crucially, though, this bizarre moment is the first time a presidential candidate has publicly insinuated that a period should be used against a woman to mark her performance down. Women duly and understandably lost their shit and began ridiculing the would-be president in equal measure. A hashtag sprung up on Twitter, Mr Trump's platform of choice, #periodsarenotaninsult as lady tweeters reacted to the businessman's comments.

> @13wildhare: 'my uterus is so disgusted @realDonaldTrump I got my period #periodsarenotaninsult.

> @rebeccagorena: '@realDonaldTrump I've been bleeding for two days and my brain is STILL working better than yours

> @kat_victorious: '@realDonaldTrump–on the third day

of my period AND still a functioning member of society! Who knew?! #periodsarenotaninsult

The most memorable protest to Trump's remarks was by an enterprising US artist, Sarah Levy, who created 'Bloody Trump'. Yes, that's right. It's everything you hoped for and didn't know was possible. A painting of Donald's face using Sarah's own menstrual blood. And very good it is too. Sarah promised at the time to donate the proceeds of the sale to charity and I caught up with her to find out what happened. She was true to her word, selling the work to the German Museum of Military History in Dresden, of all places, which has included it in a show called 'Gender and Violence'. It went for $3,000 and the proceeds were split evenly between VOZ Immigrants' Rights Centre in Portland and the Center for Economic Research and Social Change in Chicago.

Sarah was moved to go to the effort of collecting her own period in a cup and storing it in the fridge because she was so 'outraged'. Outraged that a woman's intelligence – in this case Megyn Kelly's – could be insulted because of the inference she must be menstruating.

When we look at this incident, I actually believe the greater public outrage should have been reserved for Trump's response afterwards, when trying to justify his words. He denied that he had been referring to periods: 'I was going to say, nose, and/or ears. Because that's a very common statement. [*Is it?*] Blood flowing out of somebody's nose, and remember she had great anger when she was questioning me.'

But then he followed it up with anyone who believed he *had* been talking about periods, wait for it, must be a 'deviant'.

A *deviant*. For thinking he would even mention such a dirty and disgusting thing as periods. The Cambridge Dictionary defines the word to mean a person or behaviour 'that is not usual and is generally considered to be unacceptable', and it of course often refers to sexual predilections too.

Periods are precisely the opposite of deviant behaviour. They are routine occurrences and are perfectly natural. It is only immature men who keep them in a glass box labelled 'weird'. And it is usually only those who have never had one (generally: men), who perpetuate this idea of them as unacceptable.

But Trump has form in this area of thinking that anything that a woman has going on 'down there', other than sex, is unseemly. During one of the televised Democratic debates, Hillary Clinton had to take a short break, which was presumed to be for the loo. Fair enough. We've all got to go when we need to go. Trump wasn't having any of it.

'I know where she went. It's disgusting.'

He told supporters at a rally in Michigan.

But back to the depiction of deviance around periods. I am not saying that when I read these comments, I went home and sobbed into my sanitary pad. And yet, when someone with such a large platform as Trump puts forward the idea that those people who believe he would even *mention* the word 'period' are deviants – the same word often reserved for rapists and paedophiles – it reinforces the age-old stigma that menstruation is a filthy, unusual, unmentionable occurrence. In turn, this silences and shames women; this vicious nonsense filters down from the White House down to the school playground.

Of course, many women did, and do, hit out during these moments – whether it's painted across canvas or on social media – but will their witty ripostes do much to change the majority of men's primitive attitudes towards the red stuff? Especially those men who will never have to reveal their views publicly?

I suspect not.

THE GOOD GUYS

Do not despair – there are good guys out there who champion women and their periods. Take the brilliant man who wrote to me at *The Sunday Times*, where I pen a no-nonsense advice column called 'Tough Love'. An edited version of his question was printed:

> I have a sensitive issue that I fear will make me sound insensitive. I adore my wife – she is my best friend and favourite person. We have only been married for a few

> years and life is very good – except for roughly one week a month, when her mood swings in the run-up to and during her period are awful. She becomes like a different person. The smallest thing can irritate her and start a shouting match. Even when I try to help by doing the hoovering I manage to annoy her. And then, by the end of the week, my wife is back, apologising for how she's been. I am not a sexist and know it is worse for her, but I would appreciate any survival tips.

The full version was loaded with even more loving caveats. How he really worshipped his wife and just wanted to make her life and their lives together better. And how he hoped I wouldn't mind him even daring to speak up about this issue as he really wasn't a sexist. My response was to applaud him for biting the bullet and being such a champ; to advise her to seek medical help as she could genuinely have something wrong; for them to use a digital cycle tracking app so they both knew when the storm would hit, and to have a funny safe word during the tempest – like 'frangipane' or 'kipper' – as a way of diffusing any difficult situations with a lightness of touch.

There was a powerful reaction to the piece online – it was one of the most read on the newspaper's website that week – and many emailed in their take. Women and men praised the man (whom I got the sense was quite timid but determined) for sharing his woe with such sensitivity, as opposed to squeamishness, and working *with* his wife, not against her, to improve her period.

Or take similarly sympathetic and respectful comedian and single father Gary Meikle. He made headlines in 2017 after including a series of jokes about his daughter, Ainsley, getting her first period in his standup routine. He recalls it to me down the phone:

> I brought up my daughter without the help of her mum and I was more her best pal than her dad. I'm liberal and we can talk about anything. When periods came about, it was a big time in her life and mine. She was eleven years old and suddenly officially became a woman. I came home one day and she told me 'I've got my period'. That was a big shock because up until that point I thought she was a boy. Then the two of us hit total panic stations. I said, 'don't move' and I ran to the chemist – where I told her to help me. The chemist said 'build a dungeon, don't feed her after midnight'. She gave me some paracetamol, chocolate and a hot water bottle. I was fine.

That's how Gary's period routine goes, and his daughter, now in her twenties with her own little girl, loves it. He says when he first starts the whole section on periods in his set he can feel the audience pulling back a little, but he's never had a bad reaction to it. People say to him that it's nice. Which it is, because getting your period should be treated with a lightness of touch as it's simultaneously so normal and so odd – like anything for the first time. Interestingly, in the five years he's been on the comedy circuit Gary's never seen a male comedian talk about periods – in a positive or pejorative way. 'It's a taboo thing, like jokes about the disabled,' he explains. 'The joke needs to not come from a demeaning angle or at the expense of the women as you would lose many women and men in the crowd. It would be cheap and tacky. But generally period jokes don't happen. It is a subject not talked about. I felt I could as it was a real part of my life.'

When I ask him about Donald Trump's words and men using periods as an insult his response is unequivocal. 'Men who make period jokes or remarks against women are chauvinistic arseholes. I would rip them apart. I was raised by my mum and two aunts. I needed to fight off the oestrogen. If any man tried to score cheap points about periods, I would shoot them down.'

Praise be. And men should note, this is from a man who is a successful comedian: periods can be a punchline – but not at the expense of the bleeding woman.

MEN SIGNING UP FOR PERIODS?

While Gary and my anonymous correspondent are good examples of genuine male sympathy towards bleeding women, it usually takes something else for a whole group of people to entirely change their mind about something. For instance, nobody really cares about childcare until they have children. And then you do. You *really* do.

A funny WaterAid campaign, which went viral in

2015, imagined how different life would be if all men bled every month and it made a decent attempt to spread period empathy. It was created to bring the world's attention to the horrifying fact that 1.25 billion women don't have access to a toilet during their period.

'Manpon' was the spoof tampon brand they came up with, and, of course, in this imaginary alternate universe the manpon had been designed by NASA scientists for best design and blood absorption so *serious manly sport could continue*. Periods were a sign of strength between men, not a weakness.

As part of the campaign, WaterAid conducted some research which found:

- A third of people thought that men would brag about their period
- A third believed that sports commentators would openly talk about men's cycles affecting their performance if periods were a male preserve
- Half thought that special sports sanitary products would be developed for men

But one campaign advert can only take men so far on a journey of understanding about how human life has largely been designed with men in mind.

What to do then? The only quick-fire way of ensuring men fully understand periods en masse is for them to actually *have* one, and biologically, that just isn't going to happen… Plus, why would you sign up for one if you didn't have to?

Well, in 2016, that's *exactly* what Edgar Momplaisir, an American writer and comedian, did. The twenty-seven-year-old was working at Buzzfeed when he participated in a project to see how men would cope with having a period. It involved wearing a tube connected to a pack of beetroot juice which went into the guys' pants and dripped down throughout the day, while they grappled with using pads.

When we speak, he laughs at the memory:

> For five days I wore the pack and it had a screw-like valve on it so it could periodically drip down the beet juice. There were so many things I had never thought about before. I didn't think about how if you moved a lot more might come out and it would be more uncomfortable. I didn't know you had to regularly change your pad either. This girl took me to the bathroom and told me I'd leaked. Badly. Then I had to do the whole sweater around the waist thing.
>
> You have to be so organised too. I ran out of pads one day and had ask a woman for more. I found the whole thing so exhausting and humiliating.

For Edgar, who was very game to go through the experience (despite his family and improv comedy crew not being so supportive of this weird experiment he was doing), it totally changed his perspective of women and periods.

> It made me realise that something I thought of as a

minor inconvenience before, isn't. Having a period is a big deal. It was foolish and short-sighted of me. It also made think that pads should be freely accessible. It's kind of fucked you need to go into Walgreens to buy them when you just need them there and then in the bathroom. The whole experience also made me doubt the idea of PMS when we talk about women being in a bad mood. I would be fucking pissed off dealing with this every month. The social pressure we put women under to look and be perfect – all while dealing with this and everyone staring at you the whole time.

It opened my eyes to how I was raised too. It made me realise that me and my brothers weren't sympathetic to my mom. As a man, you only have to worry biologically about how many tissues you need to clean up wet dreams. The experience made me understand how blind I was. Periods need to be normalised more. The Bible talked about women needing to be alone at this time of the month but socially I don't think we've really moved on. We still ostracise women.

Women need to be able to talk about their periods as openly as guys talk about getting wood.

Amen brother.

Now, that is the understanding I'm talking about. But remember, Edgar only had beet juice running down a tube nestled in his butt cheeks for a few days and none of the cramps or the hormones.

While the experiment worked – generating lots of clicks for Buzzfeed and changing a few blokes' minds in the process – it was very niche. Men simulating periods isn't something Buzzfeed or governments around the world are going to be rolling out across countries anytime soon.

Just like the WaterAid campaign, the experiment feeds into the fun game of imagining how periods would be treated differently if men had them. The American feminist and writer Gloria Steinem's 1978 hilarious essay describes this brilliantly. She wrote: 'Clearly, menstruation would become an enviable, worthy, masculine event. Men would brag about how long and how much. Young boys would talk about it as the envied beginning of manhood. Gifts, religious ceremonies, family dinners, and stag parties would mark the day.'

And in my favourite part of the piece: 'Street guys would invent slang ("He's a three-pad man") and "give fives" on the corner with some exchange like, "Man you lookin' *good!*" "Yeah, man, I'm on the rag!"'

People have also enjoyed playing this game on social media with the hashtag #ifmenhadperiods. To give you a flavour:

'They'd brag about the size of their tampons.'

'Tampons would be sold in the impulse buy checkout aisle right next to beef jerky.'

'A cure would have been found.'

'There would be paid time off each month.'

Certain members of the trans community have been aggrieved at this hashtag and type of discussion because they feel there is an erasing of their own experience, because trans men can, and do, have periods. It's another issue I will return to in a later chapter of the book with the input of a clever and sensitive person who describes themselves as the 'period prince'.

However, while imagining what life would be like if all men bled every month is gratifying, and a useful way to poke fun at the current status quo, ultimately it gets us nowhere. It *doesn't* stop people from thinking of periods as dirty, shameful secrets. It *doesn't* stop the leader of the free world from saying that anyone thinking he would even mention the word period is a deviant, akin to a rapist. It *doesn't* stop little boys being taught in the playground that girls ought to be teased if they do or don't want to be sexual while menstruating.

So what *does*? Pride. Shitloads of it.

That's what it always comes back to. I'm sorry to give women more jobs but it's up to us to set the tone as men simply won't. Why should they? There's no reason or incentive to do so – even those super sympathetic men who write to problem pages about their wives' raging menstrual battles or the single dads buying Tampax for their daughters.

It was during a Gay Pride march recently, amongst the heat, beauty and sheer vibrancy of the whole joyous melange of proud folk, that I truly realised the importance of pride. Pride is the opposite of shame. The *polar* opposite.

You lead by example if you have no shame. And even if the majority of men will never have a period or spend longer than a few minutes in their whole lives thinking about periods, women and girls can set the parameters and style of the conversation – with humour, truth and gritty realism in our armoury. And it's important that we do, because why should men care or change their take on it if we don't lead the way?

For far too long, it's been the sanitary product companies creating the narrative: either dressing up periods as a flowery, gentle visit from mother nature that we shoulder easily with a smile or, much worse, as something to hide, all the while bleeding clean blue liquid into tight, spotless clothing.

Now, in the digital age, we all have a platform, and we can be the real voice of experience.

Start at home. If you live with men, don't use euphemisms to describe coming on your period. Wave goodbye to 'Auntie Flo' or 'that time of the month'. You have your period. That's it. Don't spare their blushes any longer by self-censoring.

If you need to change your pad or tampon because it's uncomfortable during a family party, proudly walk to the toilet with your product of choice held confidently in your hand. And if it takes a little longer than your typical toilet trip, tell them why! New parents certainly don't have any qualms sharing in detail the forty-five minutes they spent grappling with a poo-nami in a nearby loo.

At work, if you need a sugar in your tea to get through the morning, tell your male and female colleagues why – and *don't* apologise.

We must normalise periods. And only we women can

do it as men never will. And nor should they. All we ask is that they man up to our honesty rather than shying away from the bloody reality like sniggering schoolboys.

We are in this thing called life together and period shame has been the norm for far too long.

CHAPTER FIVE

OFFICE BLOOD

'I'm not saying it was all happy-clappy. There were days when you'd just think, "Oh, my God, I've got my period and I can't get in that freezing-cold water today."... I remember standing up and saying to everyone, "Listen, if it suddenly looks like Jaws, the movie, it's my fault."'

Kate Winslet talking to Rolling Stone about filming Titanic scenes

Periods have always been used against women as weapons to keep them from furthering themselves. Sometimes it's obvious and other times, less so. We have already talked about women not being allowed to pray or enter their place of worship when bleeding. In many parts of the world, periods still prevent young women from attending school.

But in the working world, periods were still overtly used against women until relatively recently. To limit them. For instance, female pilots during the Second World War were routinely stopped from flying if they had period pain, according to Sharra L. Vostral, author of *Under Wraps: A History of Menstrual Hygiene Technology*.

Fast forward to 2011 and the ban on women becoming submariners was finally lifted – primarily because bogus science which had posited concerns that submarines' higher levels of carbon dioxide could carry risks to female health was finally seen for what it was: bullshit. But women had also been blocked from sub-sea life because of the fear men would get too horny around them in close quarters and, wait for it, the on-board toilets couldn't deal with tampons. The UK only welcomed its first female submariners in 2014. That's nearly a full hundred years after women had the right to vote.

Not so long ago, women were also advised against using sewing machines or even reading books during their periods out of fear they might 'overexert themselves', Carla Pascoe, a research fellow at Melbourne University told the *New York Times*.

Heaven forbid we pick up a book and try to take our mind off the bloody nightmare in our pants. And can you imagine trying to land a job in a society which thinks we shouldn't read or use a simple bit of sewing kit whilst we're on?

Or what about the pointless preoccupation of the NASA engineers with how many tampons Sally Ride, the first American woman in space, would need for her week-long historic mission in 1983. Their guess? A flabbergasting one hundred. For a seven-day gig. And these are some of the smartest menfolk in the world.

You get the picture of why women in the privileged West have been both covertly and overtly educated to not mention their period in a work or educational context, out of understandable fear it will be used against them. Though I must stress again at this point that women in the developing world would be lucky to have such 'first world' problems.

But in this chapter I want to convince you why period openness in the workplace is worth the potentially awful awkwardness to begin with.

It's worth remembering that even the outspoken mainstream feminist movement, in its many waves and guises, has always struggled with how to place menstruation, especially in the work context. The aim had to be to show women were as good as men in all contexts. Today, that is still the battle cry and intention of mainstream feminism. But, in the process of showing we were and are equal, anything expressly female, like periods, was de-emphasised, ignored or played down. It had to be, as the movement and its street fighters wrestled women's rights into offices, fought for equal pay bands and nonviolent marriages, as well as the opportunity to bear arms on the front line and so on. And you will know that many of these fights aren't yet won. Far from it. However, we have made huge progress on all of those fronts. As I write this, a woman runs the country I live in, another runs the most powerful country in the European Union, and sexual harassment is being called out the world over with the hashtag #MeToo.

We women are equal to men and should be treated as such. Normal, good-minded reasonable folk have always known that. Some people, though, are still yet to receive the memo. Others needed proof – of which they have now had years and years when it comes to women's records across all sections of society and different workforces.

I passionately agree with Simone De Beauvoir's take on how work can help certain women deal with their periods better, but definitely not all. The French feminist philosopher wrote in 1949, in her name-making and then scandalous book *The Second Sex*:

> The obstacle created by menstruation has often been examined. Women known for their work or activities seem to attach little importance to it: is this because they owe their success to the fact that their monthly problems are so mild? One may ask if it is not on the contrary the choice of an active and ambitious life that confers this privilege on them: the attention women pay to their ailments exacerbates them… So, a woman's health will not detract from her work when the working woman has the place she deserves in society; on the contrary, work will strongly reinforce her physical balance by keeping her from being endlessly preoccupied with it.

This is not to say any woman who can't work through her period isn't working hard enough, or that, with enough mind over matter, she can just shake off discomfort or pain. Believe me, as someone who regularly broadcasts with a hot water bottle beneath my mic desk, sat slumped to one side while I interview people, it's a real fight. I have a highly mentally occupying job keeping a three-hour live radio show going, but I am always happier being distracted at work rather than moping around the house with a hot water bottle watching ill-inducing, daytime TV, despite the Herculean effort of dragging my aching bones to the studio. Some women simply cannot do this. They have to work from home or, on occasion, lie down. Or, in the saddest cases, not work at all.

What about those women in physical jobs who stand all day packing boxes or sit at the till in an uncomfortable chair doing the same thing over and over while rushes of pain hit them? They too will also probably feel better if they *can* fight through it, and they may of course have no choice but to go in and earn their money as their line of work – contract or lack thereof – doesn't allow for working from home or semi-regular sick days. But some women sometimes just can't carry on as normal throughout the pain. Despite De Beauvoir's rallying cry, or anyone else's.

But it's one thing learning to talk to your friends, family and, if necessary, the doctor openly about your flow. It's something else to bring up your period with your boss and your work colleagues. Or that guy who sits near the loos at the office as you stagger past with your dainty special zip-up bag, waddling the final stretch to the toilet, otherwise known as 'the red mile' (too much?), having put off the soggy change for two hours too long.

And yet, even if you don't want to chat blood in the day job, it's high time we stopped *lying* about it.

I am well aware that my experience of periods are extreme because of my endometriosis. That when my pain comes on, it's gut wrenching, nausea-inducing, bone-draggingly bad. But while most women may not have a specific condition, they still often feel grim at that time of the month, require painkillers, need to access the loo more often and may encounter all sorts of other side effects, such as headaches, backache, sweatier brows and the squits. It's just a real party for your whole body during a period.

John Guillebaud, professor of reproductive health at University College London (the same guy who explained the Pope story about the Pill earlier), has actually compared period pain for some women (who don't have specific conditions or illnesses) to having a heart attack. When you stop to consider that for a moment, women are doing a remarkable job of putting up and shutting up with a monthly situation that can hit them like a tonne of bricks.

I am not arguing that women's performance at work is impacted by menstruation. Women have fought horribly hard to be allowed the right to work on the same terms and turf as men, and in most cases we absolutely manage it, month in and month out – usually with no one being any the wiser. But sometimes our lives could be made easier with small adjustments or even just the opportunity to talk about what's actually going on that day, in the same way that people do all the time at work about their migraines, back problems and even their bowels.

I return to our bowels again not just because scatological humour is a speciality of mine (though I do live with a man to whom you only have to utter the word poo and he creases up with boyish laughter), but it's a pretty sorry state of affairs when most women feel more comfortable telling their bosses they have the shits than a period. Yes, we'd rather invent a blunderbuss toilet situation than be open about our perfectly normal, perfectly healthy monthly occurrence.

Remember that study commissioned by my radio programme about women's attitudes towards periods, which I quoted at the beginning of this book? The one which found that more than half of women who confessed to experiencing period pain (91 per cent) admitted it affected their ability to work? Only 27 per cent of them told their employer the real reason they felt poorly. *27 per cent*. That means the majority of women prefer to lie. And many of those who suffered every month fibbed and told their boss they had stomach pains. They actively prefer for their boss to paint a mental picture of them pinching one out on the loo rather than riding out a healthy bleed. Other common alibis included colds, flu, headaches and medical appointments. But the very fact that faeces trumps menstrual blood – a uniquely female condition, rather than gender blind poo – is rather telling.

The cover-ups don't end there though. They can become far more elaborate, especially when women work aboard or are caught unawares by their period. Consider these two tales which, if they weren't so taboo and sad, could become excellent fodder for a Laurel and Hardy, black and white mime sketch.

One woman I interviewed in my mock radio confession booth, a softly-spoken but stoic northerner in her fifties, had never really talked to anyone but me about her period before. She'd once exchanged notes with a friend about a particular painkiller, but that had been it. For her *whole life*. She found the experience of talking about her period utterly liberating and once she started, there was no stopping her. But there was one tale which has stuck with me ever since:

> There are periods in your life when it's slightly worse than others. I don't think I've suffered as badly as some, I'm sure but it's that whole thing of once a month wearing nothing but black trousers to work, waiting for other people to leave a room because you're very sensitive about getting up and making sure nothing has happened to you.
>
> I once was working away and had to be picked up and had to get a carrier bag out to sit on in the car... Because it gets messy. Because sometimes you are suffering so

> badly, and again we don't talk about it so I'm not using the words, but your periods are so heavy that you can't guarantee you will stay clean and dry. Sometimes you go to the bathroom and you think blimey I've got a problem and my jeans are in a bit of a mess and you literally need to sit on something so it's out with the plastic carrier bag and sit on that in somebody else's car. It's not nice, and it's a good job we have a sense of humour, most of us, and we can laugh about these things afterwards. But we don't tend to have a laugh with our colleagues, we just laugh at ourselves later.

Grown women are silently sitting on plastic bags in other people's cars during business trips, praying no one will notice, and praying even harder they don't leak on said plastic bag.

Hilarious if you could josh with your work colleagues about it. But you can't. So you don't. Or consider this gem from another woman I know who works in a very male-dominated field:

> Not long ago I remember having a real 'FML' moment in the loos of Gatwick airport. I'd just got off a flight from Gibraltar after a crazy week of meetings and was rushing to more meetings in central London, when I felt that 'oh god' feeling, when you know it's coming and the only defence you have is a thin lacy pair of M&S pants.

> I made it to the loos and found a tampon, but there was some catastrophic collision of the tampon wanting to go in and a force of blood wanting to come out. There were blood spatters over my legs and all up the toilet cubicle. It looked like a slaughterhouse. I just remember feeling completely defeated and wondering why businessmen don't have to deal with pre-meeting period mop ups.

Quite. Can you imagine if she had bowled into the meeting she was understandably late for and explained her abattoir-esque warm up in the loos? The men wouldn't have listened to a single thing she'd come to say and, instead of responding with understanding laughter, they would have been utterly traumatised and she would have ended up apologising for not sparing their blushes.

Contrast this clandestine loo situation with the brilliant honesty of a woman with whom I was due to have a work meeting. She was running late because of a particular period problem and rang to tell me. The context of my working life at that point was, admittedly, as different as it could be to a male-dominated office: I ran a women's desk at a newspaper and very little was off-limits for this writer with whom I'd worked with for some time.

'Emma,' she panted down the phone, the concern in her throat audibly rising. 'I might be a little late.' I was in a cab snaking towards the office. I asked why, a little distracted by the traffic ahead. 'I'm stuck in the disabled loo with my tampon wedged inside me. I can't get it out. What do I do? It's never happened before.'

Before I could even think about what I was saying, a smile spread across my lips as I found myself respecting this woman's honesty. Then the panic of the responsibility set in. 'Bear down,' I heard myself advising in a serious tone. 'Just bear down, take your knickers and tights off, squat and bear down.'

The taxi driver perked up, riveted, thinking I was advising someone how to give birth. I was. Just not to a baby. I was not even sure what 'bear down' meant, I just knew it was the right thing for her to do – while having a good rummage. Luckily, a naked bearing down squat in the loo worked; she had remembered to lock the door, and the offending tampon was birthed and swiftly binned. Our meeting went ahead successfully, with only a few minutes extra time incurred. But I bought the tea, knowing what the poor woman had survived, minutes earlier.

Fortunately, she hadn't done the thing of forgetting she had a tampon inside in the first place and inserting another on top of it (A & E tampon stories must be a real treat) and we could carry on almost without missing a beat – save for the good laugh we had while pulling the string of our tea bags out of our steaming cups of hot water.

Again, I don't share that story to suggest that you must ring your boss or a colleague during a tampon meltdown, nor as a way of showing women being less capable during their period, but to highlight what hell sometimes occurs, why a woman might be late to a meeting or seem like she lives in the toilet at work for a few days every month.

What I'm proposing is the end of lying about periods and their potential impact on women's wellbeing at work. It's time for women to bring them out of the long shadows of the past. But I fully understand women's reticence to do so, especially those lucky ones who suffer no debilitating symptoms at all and therefore have less empathy as to why this openness in the workplace would be a positive thing (and annoyingly tend to be the most vocal in the media).

MENSTRUAL LEAVE: A BRILLIANT OR PATRONISING CONCEPT?

Which brings me onto the thorny and problematic issue of menstrual leave (minus the deafening guffaws of my embarrassed TV colleagues, this time). It's a popular policy in Asian countries like Japan, South Korea and Taiwan. I say popular, but it would be more accurate to explain that, whilst it's commonly deployed by companies, it's not regularly taken up by women who fear they will be judged for taking the few days a year they are allowed to be at home bleeding in pain and being paid to do so.

If you ask most working women what they think of the idea of menstrual or period leave, it goes down like a saturated tampon hitting the bottom of a sanitary towel bin. Miserably.

I've asked a fair few women this question while

writing this book and they have wrinkled their nose in disgust at the idea of preferential treatment, or being treated as disabled in some way for the duration of their period. I fully sympathise with this loathing of period leave. We've come this far without it – so why do we need it now?

Bex Baxter found this out to her cost as she unwittingly became the poster woman for menstrual leave in the UK in 2016 and faced the full wrath of women across the nation.

As the former people development director of Coexist, a community arts centre in Bristol, she was heavily involved in writing and shaping human resources policy at the organisation. One day, she saw a member of staff doubled over in pain as she served customers at front of house. She went over to her to enquire after her health and was batted away by the woman who said quickly and quietly 'It's just my period', before carrying on with her work.

It was at that point that Bex, who has also suffered with dysmenorrhea (aka, very painful periods) her whole life, realised 'something needed to change'.

> I realised she was ashamed and was doing what I had always done – blocking out the pain in whatever way she could and just getting on with her work. I had lived inside the bubble of the period taboo for so long I couldn't see other women struggling with the same situation. When you live inside the bubble of a taboo – it's the air we breathe – you can't see how ridiculous it is.

> I felt like I had woken up. I'd also been stuck head-down in shame about the effects of my periods every month. And even though I worked for a very understanding organisation which let me work from home or just leave early when I needed to, I still always went home feeling ashamed. I still worried about being penalised.

That was the day Bex made Coexist the first business in the UK to formally offer period leave: the opportunity for women to work flexibly around their cycles and for their colleagues to know where they were in their cycle. Bex argues convincingly that simply having a policy gave women formal permission to request working from home or to work different hours on a particularly painful day and then make up the time. Without feeling guilty, or as if they were weak. She admits that such a policy allowing fluid working patterns would not be so straightforward in larger companies, but in a small community-orientated business it was an easier sell.

And even then, I learn, it's still bedding in three years down the line, and is easier for certain teams over others. I caught up with Ruth Keenan, who presently works in Coexist's HR team.

> After a trial period, we recognised we needed to create some contingency plans for our shift workers. On the whole it works, as people do not need to use period leave that often – sometimes it's just a matter of giving people permission to have some time away from their desk in

a quieter space. And then they make up the extra time at a different point.

But there are still some sticking points – such as if you do use the policy and take time off, how does it impact others? Are they picking up your work load? Even with our relatively flexible circumstances, there are still issues. But overall, the policy has shifted the stigma surrounding period pain and discomfort – allowing women to work more optimally.

Interestingly, the policy has also been used by a man who suffers regular migraines – broadening out the concept of businesses allowing people to work at their best from just women's menstrual cycles to all people's cycles.

But away from the warm confines of her explorative workplace, it hasn't all been open minds and smiles in response to Bex's menstrual leave experiment. She felt the full ire of women across the UK when she went on ITV's *This Morning* programme to talk about the rationale behind the pilot. 'A woman rang in and said I was putting feminism back by a hundred years. It was mainly women who railed against it. Men were more empathetic! But there was a huge backlash from women who were frightened it was making us seem not as employable as men,' she recalls.

And this is where I will nail my colours to the mast. I don't believe in menstrual leave per se. I don't believe it's practical in large companies, nor do I think women need a specific policy that is expressly about periods.

It's a compelling argument when it comes to the

issue of permission: that a company's workforce needs a specific policy written so that they know they have express consent to take leave or time away from their work for a particular reason without guilt or shame. But only women who aren't directly competing with men – or those who do not care about their progression and have nothing to lose – would take up such a policy.

Intense symptoms due to period pain should be a legitimate reason to take 'sick leave', but that's all it should be. It doesn't and shouldn't need a special category, it's just sick leave. And reasons for flexible working should also include inhibiting period symptoms.

But what about the grey area in the middle? What if you are still able to perform but perhaps you just need somewhere to briefly lay your head (remember the professor who compared period pain to a heart attack?), or would benefit from a fan on your desk? How do we get to a stage where we work in environments where women feel comfortable and confident enough to talk openly about certain modifications we may need to make at that time of the month, without fear of judgement or seeming weak?

What I am in favour of is the honesty that would facilitate such changes.

Lara Owen, an academic researcher, the author of *Her Blood is Gold* and a consultant on menstruation and menopause in the workplace, is also against a policy specifically called 'menstrual leave', but does believe there needs to be some sort of formal policy to help women feel they have the right to request and make workplace modifications during their periods.

She prefers the term 'menstrual workplace policy' and envisions it as something to be developed with female employees, not imposed on them, which would help make their lives easier, especially on the first and second days of their period – which are usually the heaviest.

> It's all about the way you pose the question. If you ask women, do they want menstrual leave, most will say no. If you ask them, would you like modifications at work to make your life easier at that time of the month, they will almost definitely say yes. This could mean having somewhere to lie down for half an hour, a darker place to go away from people, the chance to leave a little earlier or come in slightly later, the ability to open the window, even cotton uniforms. Research shows that small inexpensive changes can make a major difference.

As Lara is finding, as a consultant in this space, it is a very new area with few companies having taken the plunge. Some of the shiny, forward-thinking tech firms are introducing 'wellness' policies, which can encompass anything from providing staff gym memberships to offering women the chance to freeze their eggs – yes, really. And perhaps more interestingly in the UK, much older institutions such as the police are introducing 'menopause' policies, to help women cope with their symptoms whilst continuing to work. But let's not leave periods behind – it's just as important for women to talk openly about their periods at work as it is for women to talk about their needs during the menopause.

Lara argues that there is a need for some kind of period and menopause policy as, 'Women need to have something spelled out, like maternity leave, which changed the landscape. And required all employees and management to take the matter seriously.'

While Lara and I disagree over the requirement for specific policies, what we *do* agree on is the need for women to talk freely about any problems or discomfort they are experiencing due to their period. And for their bosses to meet them halfway – without judgement.

It's just silly that women are still in a position where they are made to feel that the ideal worker body is male, and so any sign of femaleness must be concealed. Of course, I understand that at the beginning of women's mass entry into the workplace, we did anything in our power to hide any symptoms, but there is no need to do this any longer. It's time to banish our period shame to the past, whatever line of work we are in. There are enough of us women for periods – and what they do to us – to be the norm, to resist the fear of a little bit of blood being used against us. Our periods are not indicators of weakness, quite the opposite: they are badges of our humanity, a part of who we women are a few days of the week, every single month.

BRING YOUR PERIOD TO WORK DAY!

You only need to look at the recent murmurings in the world of sport – a profession dictated by a person's physical health and fitness – to see how the lingering silence around periods continues in the workplace, even at the highest level.

Fu Yuanhui, a Chinese swimmer, was hailed as a revolutionary in 2016 when she talked about swimming in the Rio Olympic Games on her period. Following the 4x100m medley relay race, in which she and her teammates came fourth, she was found crouched behind a board, hunched over in pain. When able to talk to the press, she said: 'I didn't swim well enough this time,' and apologised to her team-mates. 'It's because my period came yesterday, so I felt particularly tired – but this isn't a reason, I still didn't swim well enough.'

But don't be mistaken, women can still perform amazingly on their periods. Paula Radcliffe first broke the world record for the marathon in Chicago in 2002 in a time of two hours, seventeen minutes and eighteen seconds while fighting period cramps in the last third of the race. Can you imagine? And yet she still smashed the record by a minute and a half. A legend indeed.

Paula has spoken out about how little periods and the effects they have on elite athletes are still understood. She felt motivated to do so after she learned that British Athletics had given the runner Jessica Judd a drug called norethisterone to delay her period at the 2013 World Championships. It was a course of action Paula condemned as a leader in her field, and as someone who had tried the drugs herself to no avail. In fact, she said the drug made her feel 'a hundred times worse'. An account which tallied with Jessica's recollection of events.

Speaking to BBC Radio 5 Live about the incident, which happened when Jessica was eighteen years old, she explained:

> I had spoken to Paula Radcliffe about it and we were really trying to find ways to go around it [her period starting on race day]. I sat down with British Athletics doctors and they found this drug called norethisterone. I took that and I think it played with my hormones more. I had to risk taking that. I thought whatever happens it's going to be better than being on my period. But it still affected me, I still felt heavy legged. Especially after the race I was very emotional. I definitely wasn't myself that day.

Jessica went on to say that the effects of her menstrual cycle can be 'the difference between finishing first and last', she said she was 'comfortable' with the decision to take norethisterone as it was the 'best option at the time'.

The sporting world is a long way from figuring out its own menstrual policy. Annabel Croft, the former British women's tennis number one, has called periods the 'last taboo' in the field. But a better solution can only be getting closer as more and more athletes speak out about the challenges their periods present – instead of pretending they don't happen or burying them beneath debilitating synthetic hormones. Hopefully, now that both male and female doctors are involved in sports science, more research should come to the fore.

It's at this point I must introduce Danielle Rowley, the Scottish Labour MP, who, in June 2018, made her own bit of parliamentary history. She became the first MP to say she was on her period during a political debate in the Commons.

Danielle's period had made her slightly late to a House of Commons women and equalities debate, giving her the perfect 'in' to raise the issue of period poverty and the cost of sanitary products to the Women's Minister – an important and crazy issue we will address in short order. Huffing and puffing to catch her breath after having dashed into the chamber, she revealed her period had already cost her £25 that week in different types of sanitary products (smaller and larger tampons and pads), a variety of painkillers and spare tights and knickers following a leakage.

And while Danielle was making a bigger and very valid point about the prohibitive cost of having a period for many women, she had also just stood up in one of the stuffiest and most old-fashioned workplaces in Britain and announced she was bleeding.

Her admission and brief question to the minister duly went viral. 'If I had known it was going to get the type of coverage it did, I might have thought more about what I was going to say,' she confides to me, reflecting on the moment.

Danielle tells me that when she was younger, working in shops, she would never have brought up her period or referred to any effects she was encountering. She didn't have the confidence. As one of the youngest MPs – aged twenty-eight when she made her short period announcement – her volte face has been swift. And guess what? Nothing bad happened – save for the few usual trolls on social media who told her she wasn't fit to be an MP. She didn't lose her job – if anything, she became more popular.

Danielle also doesn't believe there needs to be a specific menstrual leave policy. Instead, she thinks flexible working should encompass a wider range of things and sick leave should also, where necessary, cover seriously debilitating period symptoms.

Regardless of where you stand on the whole debate of whether your company should institute a formal period policy, I am sure you would welcome the chance to be your whole self at work.

And that's what this is all about.

Through my journalism, I've interviewed at least two dozen people who have lived most of their lives in the closet, only to come out as gay much later on when something profound changes giving them the power to finally be honest in all areas of life. I am always struck by how no longer living a lie frees them to be fully themselves – both at work and at home. In fact, many of them report greater success in the workplace once they have come out, as they have nothing holding them back any more.

I am not for one moment comparing hiding one's sexuality to concealing one's period discomfort – but both involve lying. And lies are draining. In this day and age, we should be able to take our whole selves to work without pointless, anachronistic shame. It doesn't mean oversharing for the sake of it or being unprofessional in any way. But honesty to oneself and those around us always yields better results.

We no longer have to be homogenous robots at work. As mentioned earlier, the ideal worker body is no longer male. We should be able to be just *human*. The best employers know this.

The best bosses also lead by example. Take the bizarre case of the US congressman Sean Maloney, a Democrat, who had his expense claim for tampons used by female staff and visitors refused by the Office of Finance, which reports to the House Administration Committee (a board that is run by Republicans – a politician will always make something party political, I make no comment on that front). It was the first time the congressman's office had any expenses rejected. Bear in mind, the same budget covers buying other essential sanitary products such as loo roll and hand towels for the office toilets, too. The reason given for the rejection was: 'Tampons are not an office supply but a personal care item.'

'Tampongate', as it became known in Washington,

was soon resolved after Congressman Maloney's video about the matter attracted thousands of views and the Office of Finance backtracked. This incident happened in 2018 by the way, not last century. But it is a clear example of a boss, in this instance a man too, leading on periods, destigmatising them and making the eminently logical case for providing a basket of sanitary items in the loos.

STOP COVERING YOUR TRACKS, BE HONEST ABOUT YOUR FLOW

Believe me, I'm not arguing that women are less capable on their periods – women still save their fellow soldiers on the front line while menstruating – but we should stop pretending that they don't affect us *at all*. And we should stop covering our tracks and lying when actually, if we're on our period, we might need to work differently, even for a few minutes. We need to reach a point where talking about periods at work, and how we cope with them, is totally normal.

The woman I mentioned previously, who entered my confession box and admitted that women are 'complicit' in the conspiracy of silence at work, thinks it's purely because they fear it being used against them – that someone will say they are unfit to work because of the hormones rushing around their body. It was a powerful admission which hit the nail on the head.

That fear is understandable. Of course, some women will abuse such newfound honesty – taking lots of time off when they don't need it; using their period as an excuse to avoid undesirable tasks within their work – passing the buck to others. And some men will judge any woman openly taking leave for their period. But the majority of society is ready – and we certainly should be ready – to step into a new era where period shame is no more. Now's the time to lose the shame and *own* your periods. I think you'll be surprised how many fellow women meet you on the conversational starting blocks. And how much more confident you will feel having been honest with your boss, colleagues and yourself. Be bold and be proud. Be you. Do not let those old-fashioned attitudes put you down or belittle your candour.

If you are still nervous to blow the period taboo wide open in the workplace, call to mind these wise words from Lara Owen, who we heard from earlier:

'Any way in which women are made to feel bad about an aspect of their bodies which only females have, reproduces the female-based shame on which the patriarchy has been built. And thus keeps women down.'

Perversely, despite making you feel like horseshit, your period could be the key to unlocking your success at work – as well as giving you the opportunity to freak out some male bosses too.

Remember, a woman who is honest about her flow is not a woman to be messed with.

CHAPTER SIX

CLASSROOM BLOOD

'It's time to give periods a voice. It's time for the menstrual cycle to be heard.'

Alice, twenty-two-year-old ambassador for Endometriosis UK, petitioning the UK government to put menstrual wellbeing on the school curriculum

Let me share a little tale with you from my youth. Consider it a parable, if you like. It's certainly a story I've never forgotten:

'I think I've done it right,' my friend said, as she waddled towards us, John Wayne style, 'but I'm just not sure.' My other friend and I cocked our heads to the side and peered down. She was damn right not to be sure. The dangling string looked unusually far back as we craned our necks to gain a better view. Attempting to stifle unsisterly giggles, we realised our friend had managed to put her first tampon up her derrière.

Yup. At sixteen years old, we thought we had

counselled our pal well, at a sleepover during one of our endless school summer holidays, as she cowered around the corner of a wardrobe grappling with the machinery of menstruation.

But our shonky advice had somehow resulted in our friend ending up with a tampon up the tush – I kid you not. I can confirm said tampon was safely removed but never forgotten by all concerned.

I tell that tale as a way of illustrating how totally clueless three decently-educated young women still were about our anatomy and period products at the age of sixteen. Hell, it had taken a brilliantly clued up eleven-year-old to tell me only two years earlier that women had three holes down there, during a youth theatre rehearsal. (Yes, I was that cool.) There I was, playing this girl's mother and finding myself nodding meekly in bewildered agreement all the while thinking, 'I thought my urine and period came out of the same place.'

Again I was wrong. Utterly. *And* I failed to retain the correct information once I had it.

So how did we end up here? Despite huge scientific and technological advancements since religion ruled the roost in Western societies, there has been scant decent and accurate education in this field. That's how. And that's what I want to explore in this chapter.

Sex and relationships education (SRE) is only becoming statutory in schools in September 2020, having been delayed since it was first announced in the summer of 2018. What is the government scared of? But before you start celebrating the end of crushing ignorance, parents will still reserve the right to remove children from parts of it, should they wish to, in secondary school. And primary schools may only opt to teach lessons about relationships, rather than anything useful about bodily processes and the proper names for our body parts. So more tampons up the tush could still be on the menu up and down the country.

But to those who trot out tired arguments against schools teaching SRE, saying it should be parents who inform their children about the birds and the bees, or whatever naff way they choose to put it, decent and compulsory SRE wouldn't only be about sex. Far from it. That's the whole point. It would be about bodily fluids, menstruation products, wet dreams, our anatomy, hormones and, crucially, what's *normal* during puberty – the one question everyone has racing around their mind while growing up. There would be investment in the syllabus and teachers would be given the time and space to learn how to teach much more than putting a condom on a banana and telling kids to 'get a hand mirror and have a look and a play down there'. Well, that's how our PE teacher put it to us in one rushed Personal, Social, Health and Economic education (PSHE) class, as it was called then. My mum was thrilled when I asked for her hand mirror when I got home and nonchalantly told her why – knowing full well she'd be shocked and would try to hide her reaction.

If you can show me a parent who is adequately trained and can be bothered to do all of these unknowns justice, and a child who wants to hear this stuff from their parent, I will happily start a petition to stop this long overdue amendment to the British education system.

At this point, I must confess I am no stranger to the battle for decent sex education in our schools. In 2012, I launched a campaign about it while women's editor at *The Telegraph*. All we called for at that time was 'better sex education', not even compulsory lessons. Anything would be better, we decided, after learning to our jaw-dropping surprise, that the guidance on the subject provided to teachers hadn't been updated since 2000 (before the proper advent of the internet). A whopping twelve years. Teachers lacked the language and permission to talk confidently to their pupils about the rise of sexting, revenge porn and online porn generally – which had radically shifted young people's perspective about their bodies, sex and, again, what was normal in terms of growing up in their own skin. No teacher is going to want to freestyle around such subjects and nor should they have to.

While we succeeded on paper, in so much as the then Prime Minister David Cameron agreed in principle to an update, nothing materialised. Of course, he then hastily exited stage left post-referendum – while a succession of education secretaries entered the revolving door at the Department for Education. Rumour had it that all the women around Mr Cameron's Cabinet table wanted to seriously improve and invest in this area of schooling, while the men didn't quite share the same zeal (squeamishness abounds).

Progress was finally made in March 2017. The Prime Minister by now was Theresa May, Amber Rudd was Home Secretary, and Justine Greening was Education Secretary. All women – so the rumours may have been true. Seemingly, women needed to occupy three of the most powerful offices of state to bring the gooey awkward stuff of life onto the curriculum. Finally, this country would look to educate its children about the rollercoaster their bodies and minds were going through, as it was happening.

According to Lucy Emmerson, director of the Sex Education Forum, who is lobbying for what we will teach our teachers to teach kids, it's the timing that is so key: 'We want to make sure that children are prepared for puberty before they experience it – that is the principle we really want to make sure is at the heart of this new approach.'

She argues this so passionately because her organisation has found that a quarter of girls learn *nothing* about their periods before they started having them.

Stop and think about just how terrifying that must be. Suddenly there's blood – the most concerning of all bodily fluids – coming out of your most intimate organ and no one gave you a heads-up.

Chella Quint, the awesome period educator, comedian

and teacher, unearthed a similar story when questioning teenagers in her focus group for her PhD on periods. She discovered many mothers don't really tell their daughters about periods until their first one, in a bid to allow them to stay young for as long as possible. Ironically, the teenage boys she spoke to were told more about periods from a younger age by their mothers – instead of protecting them, they actually explained what the packets of brightly-coloured 'adult nappies' were in the bathroom cupboard.

I started my period in junior school and I, for one, would have appreciated my teachers having given me some sort of a memo about my vagina before all hell broke loose down there.

And there was no whisper about blood from my mother before that auspicious day in the toilet cubicle in House of Fraser. I once discovered my mum's sani pad stash in the middle drawer of her dresser but, despite my burning curiosity, I never dared ask for more information out of fear of being, rightly, scolded for snooping around.

Currently, if children even attend a school where SRE lessons happen, and if these lessons go anywhere near your nether regions, girls are taught about periods without mention of hormones. As Chella aptly puts it, that's like teaching maths without numbers. Boys are usually excluded. The proper names for genitals are introduced when kids are far past the point when it would have been useful to have had some language beyond ludicrously babyish names like 'nunny' and 'woo woo' – or the jarring porn script of 'pussy' and 'cock'.

Presently, in such SRE classes, bodily fluids are totally off the menu, Chella tells me with incredulity, having spent many an hour in Sheffield classrooms. Discharge? Forget it. Sperm in the context of wet dreams? Don't be daft. Blood and how much is normal during a period? You are having a laugh.

Cast your mind back. How many questions did you have racing through your hormone-addled skull when various secretions started spontaneously oozing? I know I had tonnes, and even though it would have required deft teaching to get us all through any sort of lesson with this agenda, I would have loved to have been made to feel normal – and to be equipped with this knowledge before puberty even kicked in.

But as Lucy Emmerson points out, unlike the internet and the rise of sexting, periods haven't changed for thousands of years. 'And yet even they still aren't taught about properly – without the level of shame and taboo which they have carried with them for generations. We need time on the curriculum and investment in teacher training to get this stuff right.'

Chella, a Brooklyn native, who moved to the UK because of her love of drama and Shakespeare in particular, and who began teaching SRE more than a decade ago, couldn't agree more. She took smashing the period taboo to a whole other level by creating comedy about the red stuff in her spare time on stage in stand-up routines and in her magazine, *Adventures in Menstruating*.

And yet even she was concerned that her school in Sheffield wouldn't approve of her 'secret double life' as a period stand-up comic – not your most regular side gig to a teaching career (although one she highly recommends). Despite trying to smash shame, she worried her employers would think she was 'dirty' in some way, and that her reputation would be sullied. Thankfully for her, they didn't mind a jot – especially as she then went onto deliver a popular TEDx talk on how tampon companies use shame to sell products, and a lecture on the same subject at the London School of Economics.

Chella has a simple compelling theory: that previously it was religion and culture which inculcated the shame of women bleeding, now it's capitalism. It's the period brands telling women to 'shh' and 'whisper' about their period. That they smell and so we need flowery chemical scents down there. The companies create adverts which show a thick blue liquid instead of dark viscous blood, and which feature women skipping about in skin-tight white pants while happily eating a yoghurt, smiling breezily as their body easily takes the waves of pain and blood raining down into their teeny tiny pants.

That's the way it was decided to market periods to make money – like they're a gentle but smelly secret and something that beautiful women handle without batting an eyelid.

Such a powerful narrative has proved contagious.

The same approach has wormed its way into our societies the world over and into our schools too. It can't be right that when teaching about something so normal and frequent as a monthly bleed, boys are often sent out of class and the person conducting the lesson is generally someone different to the normal teacher – like the school nurse. As Chella puts it, the only other subject that is treated in such hushed tones, demanding the intervention of someone other than your regular teachers, is child protection.

Periods should not be existing in the same bracket as a potential crime.

Of course, this is why the drama has built up around periods. Schools don't tend to progressively teach about them from a young age. They hit you with one biggie, often too late, which contributes to this narrative of building them up into some sort of terrifying occurrence, taken out of all context. It already seems like a horror film down there – why do we need to up the ante even more?

Chella is trying to stem the tide by aiming to make her adopted home town of Sheffield the first 'period positive' place in the UK. This would mean all schools would be vetted on how healthily they deal with periods – like a Fair Trade stamp of approval, but about the red stuff.

Being period positive is a bloody brilliant concept – you don't have to love periods but the way you talk about them should be positive and straightforward. That's it. No hushed tones and no shame. The measures Chella wants to see in place across all schools include:

letting people go to the toilet more easily during class and exams without grand inquisitions; small pedal bins in the stalls of all toilets, a range of free menstrual products being made visibly available around the school, from the library to the examination hall to toilets themselves; boys included in menstrual education, and all teachers being trained in how to respond if they are asked for a tampon or pad. This is on top of the hope for a much improved curriculum where periods are talked about from a young age, not introduced after the event as giggling boys are ushered out of the room for a bit of extra footie time.

The worst thing about Chella's stonkingly brilliant vision is how utterly simple it is yet how far off it still is for most schools – which, in another vicious cycle, are run by folk unwittingly and benignly infected with the shame and giggles of how periods were or weren't taught to them.

So, if we cast our minds to the fundamental, overarching question, 'why don't men and women understand periods?' the answer is achingly obvious. We need to decently educate girls and boys in school without shame about the realities of their bodies, including periods.

Then maybe, just maybe, we won't have another girl sticking a tampon up her tush. Period.

CHAPTER SEVEN

'Women belong in places where decisions are being made... It shouldn't be that women are the exception.'

Ruth Bader Ginsburg, a US Supreme Court Justice

Considering the paltry state of period education on offer in the UK, it's hardly surprising that the top debating chamber in the land, the House of Commons, can't cope with this subject either – or at least, hasn't been able to until very recently. (And even then, parliamentary business juddered to an awkward halt while terrible 'period jokes' ensued.)

The fiasco of menstrual products being deemed 'luxury items' for tax purposes is one of the starkest examples of what happens when you don't have women in positions of power, and boys, who become powerful men, aren't properly educated.

Let's find out how tampon tax came about in the first place – and how it was finally defeated.

First of all, it took until 2016 for the word vagina to be said aloud in the hallowed chamber. Yes, that's right. You know that doorway through which most of us spring forth? Utterly unmentionable in Parliament it seems.

The Great Vagina Silence, as I like to call it, was shattered by a certain Paula Sherriff, Labour MP for Dewsbury, on a historic day in March 2016 when she dared to ask David Cameron about plans by the government to accept Labour's tampon tax amendment to the budget, and whether the 'Vagina Added Tax was going to be consigned to history'.

The then Prime Minister, never one to miss the opportunity for a schoolboy joke, duly replied, with a wry smile, that explaining 'sanitary products' and tax to his European counterparts (because when we joined the EU in 1973 it was deemed perfectly acceptable to tax menstrual products as a 'luxury' and a popular UK petition you are about to learn more about had forced his hand) was a day that would stay with him. Roars of braying laughter bellowed out from his MP colleagues as he sat down, seemingly very pleased with his ability to even mention sanitary towels in the Commons.

So, this is how the tampon tax debacle began.

I've never cared much about tax before. I'm happy

to pay it when it automatically leaves my pay cheque and I don't have to think about it again, but engage with the stuff? No thanks. I'm baffled when I go to America and see sales tax added to everything after the event, making all purchases seem far more expensive. I know I'm not alone. But ignorance is rarely bliss. And I had no idea about the tampon tax or 'vagina added tax' until university student Laura Coryton became rightly and properly aggrieved about it and launched the 'Stop Taxing Periods' campaign in 2014. The online petition amassed 320,000 signatures, eventually Number 10 paid attention to it and hundreds of thousands of women slowly awoke to the fact that crocodile meat isn't deemed a luxury in this country and consequently isn't taxed – but bizarrely tampons are. The skewed logic which underpinned this wrong-headed decision to tax sanitary towels and tampons in the first place is because we pay tax on items deemed non-essential and are therefore considered a luxury as opposed to an absolute necessity. You find me a menstruating woman who would consider a tampon or a sani pad anything other than a total necessity and I solemnly swear to stop eating my favourite food of chips until the end of my days. That's how confident I am – I'm willing to stake what would be my last meal on death row forevermore. No woman ever thought 'oooh you know what? I am going to indulge myself in a little luxury this month and buy a box of tampons! What a naughty treat'. Instead the narrative in the supermarket is more akin to buying loo roll: 'Shit I've nearly run out. Must buy more now. Right now.'

Paula Sherriff threw her weight behind Laura's campaign and took over the political side to help push the government into change. She was successful. In the same month that she first uttered the word 'vagina' in the House of Commons, Parliament accepted her proposed amendment that would end tampon tax once and for all in the UK.

BUT KEEP THE BLOODY MARYS ON ICE FOR NOW...

Frustratingly, Brexit has thrown another spanner in the works. That's right, because nobody quite knows when the UK will actually leave the EU (or just keep endlessly transitioning), we don't know when the tampon tax will end. The government has promised that when we are out, sanitary products will finally be zero-rated – as supposedly, we will have control of our borders and tax codes. While we remain in the European Union, we have to obey the rules set for all 27 countries and tampon tax is a constant throughout the union. As we have heard endlessly throughout Brexit debates – it cannot be one rule for the UK and another for all other states. The same sadly applies to the bogus tax law deeming tampons a fricking luxury. For many Remainers, including Laura, the abolition of the tampon tax has become a tiny Brexit silver lining.

Never one to miss a political opportunity, the former leader of UKIP, Nigel Farage, who had (unsurprisingly) never taken a public interest in menstruation before, seized upon Laura's tampon tax campaign in the run up to the EU referendum in 2016, and utilised it as another example of the ludicrous rules imposed on the UK by the grey bureaucrats. Hijacking women's uteruses for political gain? Nigel knew he was onto a winner. Surely, there could be no other reason for his feminism suddenly coming to the fore at that precise political moment?

Until we transition into full-fat Brexit, the tampon tax fund (launched in 2015 by the then Chancellor of the Exchequer, George Osbourne, with all the gusto of a deflated bag of quavers) is the sticking plaster. (The sight of George saying the word 'tampon' is still one of Laura's favourite memories.) It works like this: every time we buy women's sanitary products, the VAT charged goes straight into this new funding pot, which in turn gives out roughly £15 million a year to women's health and support charities, such as those benefitting victims of domestic abuse and rape.

So, if you think about it, women in Britain are now paying tax twice – for being *women*.

We must spend our hard-earned cash on menstrual products we cannot do without, and that same money is spent on women's charities that arguably the government should be funding the hell out of anyway – crazy, or what?

Perhaps all of this nonsense could have been avoided if we were educated to know that periods are just another bodily process, not a luxury only women are lucky enough to experience. I digress. As politicians are so annoyingly fond of saying: 'We are where we are.'

Paula Sherriff told me that she felt both 'mischievous' and 'empowered' saying the word 'vagina' that momentous day in Parliament. 'If women and girls could see me talking about periods and body parts in the House of Commons, then I hoped they would feel they could talk more openly anywhere about these taboo things. It's like mental health – we need to talk about periods,' she explains.

And yet, she still had to deal with some backlash afterwards. 'Later that week, an older male Tory MP, who I won't name, came up to me and asked "Was it absolutely necessary to use that word in the chamber?" I inquired which word he was referring to. And he managed to reply "vagina" and looked as if he needed smelling salts.'

Yes, a man elected to represent the people of his constituency (of which I suspect 50 per cent will have been women), couldn't deal with the right and proper word for a woman's genitalia. Online, the reaction isn't much better. Paula received the customary sexist abuse for daring to raise 'wimmins' issues and became known as 'Tampon Woman'. Paula rightly sees her new moniker as a badge of pride, though.

Sixteen years before Paula, another female former Labour MP, Dawn Primarolo, travelled down this lonely and bizarre road. In 2000, she succeeded in lobbying the then Chancellor, Gordon Brown, to slash the tampon tax rate from the standard 17.5 per cent to 5 per cent. By all accounts, it wasn't easy. At the same time, there was a separate lobby for cutting tax on sunscreen – and eventually Dawn had to argue that you could choose to sit in the shade but you couldn't opt out of a period.

But the bigger point I would like to make here, is that the general public didn't even hear about this at the time – well, certainly not from the Chancellor himself. According to Damian McBride, the former Labour spin doctor, famously awkward Gordon Brown didn't want to utter the word 'tampon' at the dispatch box while delivering his Very Important and Very Serious Budget – so he simply didn't mention the tax cut at all.

Indeed. A fully-grown man, in the twenty-first century, couldn't and *wouldn't* say 'tampon'. Let that percolate in your mind for a minute.

Of course, we can laugh about it, in the way only the British do. Let's take it on the chin and make people feel foolish through our chortling. The late Labour MP and former Northern Ireland Secretary Mo Mowlam was the master – she loved telling ministerial colleagues that she instructed her security detail to buy tampons for her.

But there is a serious point here. Yes, in 1973 when the

tampon tax was quietly ushered in, Margaret Thatcher may have been the only woman in a Cabinet position, but (love or loathe her) most women would understand why she didn't go into bat for periods at that time, if she even had the opportunity. When you are in a minority, you don't want to seem like you are kicking up a stink about something which nobody else fully understands, or at least at that time felt like they could talk about. Obviously, I wish she had. It would have been awesome, brave and avoided a lot of nonsense.

But Britain is not alone in taxing women for menstruating. As we know, it happens across the EU, but the same policies are in effect all over the world. Stateside, forty out of the fifty states do it. That's right, in Texas cowboy boots are deemed essential items but sanitary pads aren't. When asked about the tampon tax by a YouTuber while he was still President in 2016, Barack Obama was momentarily blindsided. Yes, he scrabbled together a sensible answer about needing more women at the top tables making decisions, but it was clear he didn't even know this levy was a thing. Barack Obama only being the leader of the free world. At the time of writing this book, a mere ten states have abolished a sales tax on tampons after outraged women campaigned for change.

The good news is that instead of a global tampon tax being one great, big, hateful conspiracy against women, it seems to be underpinned by something far more pathetic and juvenile. It comes down to three factors: there were no women leading our countries when this nonsense came about; no one could say the words tampon or vagina; and women's experiences were not spoken about – by women or by anyone else.

Jennifer Weiss Wolf, lawyer, author of *Periods Gone Public* and one of the lead campaigners trying to quash the US tampon tax, thinks that it points towards the bigger issue of a policy landscape that has excluded women's needs for generations:

> People are surprised when I tell them that our [American] health and safety rules mandate by law that there is soap and toilet roll in all bathrooms – but tampons aren't included. They are surprised when they learn how our food and drug administration classify menstrual products, meaning American consumers aren't allowed by law to learn what ingredients go into our tampons and sanitary towels.

Jennifer knows menstruation isn't as pressing as national security or healthcare, but what she nails is that highlighting major oversights like the tampon tax in an opportunity to expose 'what happens when society doesn't treat women's bodies as the norm'. As she puts it, society is still in the position where most of the time 'the ones who don't menstruate make the rules'. And it's only when women realise that there's a law mandating soap and toilet roll – but not sanitary products – they suddenly see how mad that is and feel a mixture of surprise and outrage.

Today in the UK, there's an All Party Parliamentary

Group (a committee of MPs from all political persuasions) looking at how best to help those from ethnic minority backgrounds talk more about their menstrual wellbeing, the effects of menopause in the workplace and how other obstetric complaints can be diagnosed earlier and receive more funding. It's a brilliant step in the right direction, but we still have a long way to go.

Paula Sherriff sees the tampon tax as the cotton wedge (sorry, couldn't resist) to open even bigger discussions around women's health. It's bitter experience that fuels our 'Tampon Woman', because since the age of thirteen, she has suffered agonizing periods:

> Even though I was very close to my mum, I have never spoken to her about my period problems or sex. Or anything of that nature. I suffered in silence for years. And when I did start going to the doctors, they told me to deal with it and gave me lots of contraceptive choices. There was a reluctance by them to do any further investigation until finally I was referred to a gynaecologist, who's a monster. He didn't believe me about my pain and made me feel like it was all in my head. He suggested I was imagining my issues. Eventually I was referred to another doctor and he was amazing. He arranged for me to have proper investigations and my options were to do nothing, have a hysterectomy – which I considered but I wanted children – or have a medical menopause induced.

Aged 37, Paula opted for the medical menopause. Strikingly, she doesn't even remember what her condition is called when we talk. She says that whatever it is, it mimics endometriosis, the painful disorder I have and know all too well. She still has periods but is coping much better now, despite having to accept she will probably never have children. It's something she says with sadness in her voice. Paula also admits to me that her painful periods really affected her mental health for twenty years and made her feel like she didn't want to leave the house. This, from a straight-talking northern woman, is a sobering admission – but she is not alone.

Luckily, Paula has the confidence – and now the platform – to do something about the lingering silence. Prior to coming into politics, Paula worked in a hospital and there she was still squeamish about speaking to anyone about her symptoms because, as she so punchily puts it, she didn't want doctors in the same place seeing her fanny – before then telling herself off for using 'fanny' instead of 'vagina'.

If Paula couldn't bring herself to sort her fanny out, as a self-assured grown woman, imagine how shy, awkward schoolgirls deal with heavy painful periods – especially those without the financial means or language to express what they are going through.

FIND YOUR FANNY VOICE

Fortunately for all of us, as we know, Paula did eventually find her voice on fannies – and in the most powerful room in the land, to boot. Yes, there may have been muffled laughter and probably a good helping of old-fashioned outrage in the House of Commons as she made her case, but the milestone moments of Paula, and other MPs like her, are gradually helping to redress the weird political wrongs of the past and finally bringing periods out into the open.

But let us not forget, despite such progress, women are still being taxed for their periods all over the world, tampons are still deemed luxury items in Western countries and male chancellors (the UK has never had a female one) still look as if they might just choke at the mere mention of a tampon.

Once we are able to educate girls and boys properly about periods, it should naturally follow that we have a society filled with adults who will never again tax women for being women, disbelieve them in doctor's surgeries or wrinkle their noses at them in political debating chambers.

But the change must start now with those *already* in power. We can't sit idly over the years to come, waiting patiently for this information to filter down – we owe it to the next generation to start this wave of change now.

So, find your fanny voice, and fight for its rights loud and proud.

CHAPTER EIGHT

POOR BLOOD

'No one knew so no one helped.'

Rachel Krengel, a young mother of two living in London, who had no money to buy tampons or pads

That's right, period poverty is right here on your doorstep. Not just in India, or Africa, but here in Britain. In 2017, Plan UK conducted an online study of 1,000 girls and young women, aged between fourteen and twenty-one, which revealed the extent of period poverty in the UK. Some of the results, which I have included below, might – and should – shock you:

- 1 in 10 girls are *unable* to afford sanitary wear
- 1 in 7 girls *struggle* to afford sanitary wear
- 1 in 7 girls have had to ask to *borrow* sanitary wear from a friend due to affordability issues
- 1 in 5 girls have changed to a *less suitable* sanitary wear due to affordability issues

And it doesn't stop there, because the study throws up further statistics that demonstrate the taboo and stigma surrounding periods and menstruation:

- Half (yes, *half*) of girls aged 14–21 in the UK are embarrassed by their periods
- 1 in 7 girls admitted that they didn't know what was happening when they started their period
- Only 1 in 5 girls feel comfortable discussing their period with a teacher
- Almost three quarters of girls admitted they have felt embarrassed wearing sanitary products
- 1 in 10 had been asked not to talk about their periods in front of their mother or father (shocking, isn't it?)

And, wait for it…

Half of girls have missed an *entire* day of school because of their period.

It's an important and hugely troubling study – and yet, most people don't know what the term 'period poverty' even means. As we know, the word 'period' isn't commonly used in day-to-day chit-chat, so the idea of attaching the word 'poverty' to it just doesn't compute for most people. Put simply, it means not having enough money to buy the basic products to manage having a period. And it's certainly not a problem limited to the UK, a wealthy Western country that should be much further along this road.

Let's also remind ourselves of the menstrual plight and stigma faced by our sisters in other continents, through the words of Meghan Markle, on 8 March 2017, after she returned from a trip to Delhi and Mumbai with the charity World Vision to meet girls 'directly impacted by the stigmatisation of menstrual health and to learn how it hinders girls' education':

> 113 million adolescent girls between the ages of 12 and 14 in India alone are at risk of dropping out of school because of the stigma surrounding menstrual health. During my time in the field, many girls shared that they feel embarrassed to go to school during their periods, ill-equipped with rags instead of pads, unable to participate in sports, and without bathrooms available to care for themselves, they often opt to drop out of school entirely.

While Meghan was visiting the slum communities of Mumbai, she shadowed women working in a micro-finance organisation which made sanitary towels and sold them within the community. She wrote about this organisation in a piece for *Time* magazine:

> The namesake of the organisation, Myna Mahila Foundation, refers to a chatty bird ('myna') and 'mahila' meaning woman. The name echoes the undercurrent of this issue: we need to speak about it, to be 'chatty' about it.
>
> 97 per cent of the employees of Myna Mahila live and work within the slums, creating a system which, as

> Nobel Peace prize nominee Dr Jockin Arputham shared with me, is the key to breaking the cycle of poverty and allowing access to education. In addition, the women's work opens the dialogue of menstrual hygiene in their homes, liberating them from silent suffering, and equipping their daughters to attend school.

Amen Duchess.

We all need to be chatty birds and end the silence around periods.

Fast forward a year, and as Meghan prepared to walk herself down the aisle of St George's Chapel in Windsor to become the Duchess of Sussex, and marry into the most famous family in the world, she chose – alongside her husband-to-be – to select the Myna Mahila Foundation in India to receive financial contributions in lieu of wedding gifts. The organisation was the only non-British charity on the list, but of course, as Plan UK's study revealed, period poverty is still very much rife in the UK.

Heartbreakingly, two fifths of girls in the UK have had to use makeshift sanitary wear (such as folded loo roll) because they struggled to afford a pad or tampon. And 7 per cent of this group admitted to using socks, newspaper or other bits of fabric to absorb the blood. Let that sink in, young girls in Britain are using socks to soak up their menstrual blood. Can you imagine how mortifying, how humiliating that must feel?

Insights into the reality of living in period poverty are rare, but another glimpse comes via the writer Sali

Hughes. A secondary school teacher recently confided to her that 'some ordinarily well-behaved girls now rebel, acting loudly and antisocially, in the hope of being marched out of class and into the privacy of the corridor, where they can apologise and ask "Miss" discreetly if there's any chance of a towel'. It's gut-wrenching to read. And both male and female teachers have told Sali that they have been stocking up on sanitary towels for their poorest female pupils for the past few years.

Whilst we must applaud the actions of these teachers who are going above and beyond for their students, it should also go without saying that teachers should not be using their own salaries to supply something that should be a basic human right.

Enter the increasingly cutting-edge Girl Guides who, in a bid to raise awareness about this clandestine issue, have introduced a new woven medallion for their members to earn: the 'End Period Poverty' badge. In creating such an explicit badge, the Girl Guides hope to challenge its youth to engage head on with the issue of period poverty, to seek out their classmates who may be struggling in silence and to help them with sensitivity.

When the badge launched in May 2018, there was almost inevitable pushback with critics saying the Guides had gone too far in making a badge about something which should be perfectly normal and unremarkable. Of *course,* periods should be just that. But they aren't for so many people – especially the girls and women who are spending an entire week of every month unwillingly free bleeding into their clothes and panicking about where their next makeshift absorbent piece of fabric is coming from. Until attitudes change and sanitary towels and tampons are made freely available to those in need, (and ideally all women) the badge stays.

(Plus, who wouldn't want a badge featuring a purple heart filled with every kind of sanitary product, from tampons to Mooncups, sewn onto their clothing? I know I would. But alas – I wasn't even a Brownie).

Period poverty in a rich, supposedly enlightened country like the UK, is a totally solvable problem. And yet it took the actions of a schoolgirl to come up with a solution within our education system...

Amika George, who lives in London, founded the Free Periods movement in 2017, when she learned that there were girls across the UK skipping school because of period poverty. Her idea was simple and logical: that girls on free school meals should also be given access to free sanitary products.

But thousands of signatures later, and countless celebrity endorsements from Daisy Lowe to Suki Waterhouse, she still hadn't heard from Number 10.

'It's hard to do my A-levels and keep trying to get a response from Number 10,' she told me, sighing on the phone in a snatched moment between revision and exams, with a candour and busyness that makes me laugh when I think back to what I was using my time for at her age (fighting with my parents about using the phone, watching endless *Sex and the City* episodes and doing unnecessarily late night revision).

'Girls have emailed to tell me that they can't ask their parents for period products as they know they are already struggling to put food on the table and they don't want to put them under any more pressure. Plus, they are embarrassed to talk about it,' Amika explained. 'It must be very hard as their mums will know what their daughters are going through but they've got to make choices. Quite early on I heard about a girl who would carefully comb through the creases in the sofa to find change to see if she could buy some sanitary towels.'

In December 2017, Amika led a protest of 2,000 people outside 10 Downing Street, with talks from Adwoa Aboah, Tanya Burr, Aisling Bea, Daisy Lowe and Jess Phillips MP. Two months later, the government pledged £1.5 million of the tampon tax fund to the sexual health charity Brook and Plan UK.

Both organisations are now looking at trialling a P-card scheme, similar to an existing C-card scheme (a way of getting free condoms to young people and educating them about safe sex at the same time). At first, the P-card will target places where young people are experiencing housing issues, working with the homeless charities Centrepoint and The Foyer Federation, but the ambition is also to work with local authorities to get it into schools. It will be a long and arduous road, and will require constant funding.

One year later, on 9 March 2019, now with 271,000 signatures on her petition, Amika finally heard the news that she (and women across the nation) had been waiting for: Chancellor Philip Hammond pledged to end period poverty in schools by funding a scheme to make free sanitary products available to all those in primary and secondary school and colleges from early 2020 in England.

It's a monumental turning point in period history, especially when you consider recent chancellors couldn't even stomach saying the word 'tampon', let alone budgeting for them.

Rightly, and ever the campaigner, Amika wants this fund to be ring-fenced so it is extra money, specifically for sanitary products, meaning financially stretched schools don't face further hits to their school budgets.

PERIOD POVERTY BEYOND THE SCHOOL GATES

But it's important to remember that period poverty isn't just a problem for schoolgirls. Women of all ages and backgrounds struggle to afford sanitary products in the UK and beyond. Behind every schoolgirl trying to find a pad, is usually a mother who definitely doesn't have any period gear herself – because if she did, she would have given her last pack to her daughter. She is often the one making tough family choices about what her meagre budget goes on that week: hot evening meals or sanitary products.

Rachel Krengel, a mother of two and an activist for Fourth Wave LFA (creators of the #PeriodPotential campaign), is one woman who is refusing to be silent about her experience of period poverty in Britain. One month, her benefit payment was delayed. 'We fell into this awful spiral of poverty and debt,' she told CNN. 'One of the things that fell by the wayside was my ability to buy enough menstrual products to get me through the month... I never talked to anyone at the time. I never told my partner it was an issue.'

That's right, she chose to shoulder the grim reality of no sanitary products *alone*, something that's just as serious as not having toilet roll, rather than tell her partner. And so, just like the young girls skipping school out of fear of leaking without any sanitary products, Rachel stopped going out when she got her period – choosing instead to leak into her stained jeans without the worry of anyone noticing.

Wrestling with those sorts of dilemmas month in, month out, will take its toll mentally and emotionally. Especially at a time which can already be a hormonal rollercoaster for many women. So, period poverty doesn't just pose a risk to a woman's physical health (there is a greater risk of contracting thrush or a UTI when using makeshift towels or products for longer than they should be), it can also wreak havoc with mental health.

Dr Sarah Simons, a junior doctor in the UK, paints a vivid picture of how dark it can become in a blogpost on the subject:

> I've encountered several people who have resorted to shoplifting in order to get pads or tampons they so desperately need every month. Some of them have ended up with a criminal record for doing so, which has in turn had a knock-on effect on relationships and their ability to get a job.

Digest that. Women really are getting criminal records because of their periods, which then further impacts their ability to earn money and land a steady job.

It's a cruel, mad and totally unnecessary situation.

Luckily, there are now several organisations that have sprung up to help ensure that free sanitary towels and tampons reach those women without funds and to stop women feeling so desperate they need to steal.

Bloody Good Period, set up by fellow Mancunian and all round good egg Gabby Edlin, specifically targets female asylum seekers. Gabby founded the mission after she volunteered at a drop-in centre for asylum seekers and refugees in north London and realised sanitary products were classed as 'emergency items' – meaning they weren't readily available to all women visiting the organisation. Women who have escaped from the terrors of their own country deserve a decent, stress-free and clean period. It's the least we can do while providing them sanctuary.

Similarly, organisations such as Period Box are trying to ensure as many food banks as possible across the UK offer sanitary towels as a basic essential. Again, women are often too embarrassed to ask if a food bank has these products – and typically they haven't been stocked in the past. But at least there is now a concerted effort to solicit donations for them and for the pads to be proudly displayed next to the tins of baked beans. Both are vital.

Another group of women in even more need when it comes to period supplies are the most invisible of all: the homeless. If it's bad enough free bleeding, aka leaking, all over your clothes in the confines of a house with easy access to a toilet and a sink, imagine how horrendous it would be doing it on the street.

Crisis, the national homelessness charity, estimates women make up 26 per cent of the UK's known homeless population.

But as women are a big portion of what is known as the 'hidden homeless' those people charities can't easily locate as they literally hide in friends' or family's homes kipping on sofas, and don't seek any formal support, it's believed that this percentage is actually far higher. These women's needs, especially personal ones, remain concealed too.

Fortunately, campaigns and groups have sprung up to try and help the homeless women who can be found and still aren't very well served on the personal front. The Homeless Period is one of these, and is petitioning the government to provide shelters with an allowance to buy sanitary products the same way it offers free condoms. The campaign group also encourages members of the public to donate pads and tampons to their local shelters and foodbanks. The Homeless Period's strapline says it all: 'It doesn't bear thinking about… and that's the problem.' It really doesn't.

And then there's FlowAid, set up by Hayley Smith, who by day works in PR, and spends the rest of her time trying to help homeless women access sanitary products and restore some of their dignity. Suffering herself with particularly heavy periods, Hayley spoke to homeless women about what it was really like menstruating on the streets, after she discovered there were no statistics or any research on the subject.

'What I found out was that tampons were like gold dust on the streets,' Hayley told the former women's lifestyle website The Pool. Any money they had was spent on food, rather than sanitary products, and, as such, women were dangerously leaving tampons inside themselves, or even re-using them – causing a much greater risk of toxic shock syndrome and urinary tract infections.

But the analogy has an even darker side, as Hayley soon discovered when she visited homeless shelters. When volunteers placed a newly-filled box of sanitary products on the counter for women to take, it was homeless *men* stocking up. Why? 'For controlling women on the streets, to use them [sanitary products] to get drugs, cigarettes, sex...'

Leverage.

Tampons, it turns out, really are gold dust on the streets.

That's why part of Hayley's work, on top of driving donations of such products into shelters, is training those who work in the shelters how to safely and privately distribute tampons to the women in need. FlowAid also goes out onto the front line and hands out sanitary towels directly to women on the streets.

All of the work going on to try and tackle period poverty points in one direction: make sanitary products free for those in need.

But what about if they became free for all women – regardless of their bank balance?

The eyes of the world should turn to Scotland. The Scottish government was the first in the world to provide free sanitary products to all of its students. That's all 395,000 school pupils and students across colleges and universities. Every menstruating student, regardless of age or background, now has access to free pads and tampons galore in a bid to 'banish the scourge of period poverty'.

The move came after research in 2017 by the Women for Independence group showed that nearly one in five women in Scotland had endured period poverty – having a significant impact on their wellbeing, health and hygiene levels.

The historic initiative began in the Autumn of 2018, costing a punchy £5.2 million. While the main aim of the project is to ensure no young people miss out on education because they can't afford a tampon, there's also a hope it will reduce the taboo around periods generally. As Scottish councillor Alison Evison says of the scheme: 'It will also contribute to a more open conversation and reducing the unnecessary stigma associated with periods.'

North Ayrshire council picked up the baton in August 2018 and believes it is the first local authority in the UK to provide free sanitary products in all public buildings

– which includes community centres, libraries and public offices.

Joe Cullinane, the Scottish Labour leader of the council, says the councillors were inspired by a local school scheme providing free pads to schoolgirls. 'When you've got that kind of momentum in schools, you think that periods don't stop at the school gates, so what about their sister or mother?' Cullinane told *The Guardian*. 'How can we normalise this even further, so that it's like providing toilet paper or hand wash?'

He estimated the cost to be £40,000 – the vast amount of which will be spent on setting up vending machines in the women's toilets. Other businesses and organisations have also followed suit – including Celtic football club.

Monica Lennon, a Labour Member of the Scottish Parliament, wants to take Scotland's pioneering approach even further. She has introduced a Member's Bill to launch a universal scheme so that sanitary towels and tampons are free *everywhere* in the region.

Can you imagine the cultural and financial shift if periods were suddenly free and not taboo? It would turn a whole industry (one you are about to read a lot more about in the next chapter) on its head. Not to mention removing the stress from the lives of many struggling women. Periods cost a woman in the West between £5,000 and £18,000 over a lifetime. The lower figure covers the basics and the higher includes new knickers, painkillers, chocolate and other indulgences to help the time pass more enjoyably.

As Scotland tries to explode the entire way we 'do'

periods, it is certainly the nation to watch. Will the taxpayers protest? Will the period companies be forced into doing cheaper bulk deals with the Scottish government? Will it be a short-lived trial? Or will it finally smash the whole damn stupid taboo and make period poverty history and spread like wildfire?

I know what I'm hoping for.

But it isn't just about the practicalities of providing free tampons. And shushing the whole thing up. Yes, we need to make sure that all women have easy access to tampons and pads, but ending period poverty is about ending *shame*.

It's about no more girls missing school because of their period – and then lying about it. It's about stopping girls needing to use a sock in their knickers to soak up their flow whilst living in one of the wealthiest countries in the world. It's about women no longer getting a criminal record in the process of concealing a natural bodily process.

Politicians *must* sit up and listen to this shocking reality: period poverty right here in the UK is trapping adult women in their homes and robbing young girls of their education.

Repeat after me: It. Does. Not. Need. To. Be. Like. This.

CHAPTER NINE

RICH BLOOD

'Do you think Apple would have released its much anticipated "Health" kit without the ability to track periods if there'd been a woman high-up in the organisation? I don't.'

Baroness Martha Lane Fox, 2015, founder of lastminute.com who now sits on the board of Twitter

The basic design of modern sanitary products hasn't really changed since the 1930s, when they were introduced. The giants of Bodyform, Kotex and Always have remained queen. A little moderation has been made over the years, apparently making pads more tailored to our differing labia sizes, while tampons have become smaller and applicators easier to use. But we still waddle around using scarily similar sorts of absorption aids to our great grandmothers.

Throughout history, we've had a blind spot when it comes to periods – and businesses are no exception.

But even more bizarrely, even the companies that are part of this very industry seem to have a collective squeamishness about our monthly bleeds, and they're the ones dictating the tone of our periods. It's ludicrous, especially when you collate these corporate misdemeanours in one place and consider the size of the market. And I'm about to show you how this has a major ripple effect on the rest of society.

It's time to point out the madness of the status quo, then we can see what we can *all* do differently day-to-day, armed with that awareness.

Until now, you might have thought that religion was powerful in influencing our daily lives. Try big business. We have all been brainwashed on periods by it, whether we want to admit it or not. Because when most people are trying to avoid talking about menstruation, the majority of the conversation has been left to those who *profit* from women's blood. Yes, they are the only ones who are financially incentivised to have this chat – but you won't be surprised to learn that I think that it's time we should change that. You and me.

'Live fearless.' 'Rewrite the rules.' 'Unstoppable.'

Do any of these super-duper positive words describe how you feel when your period arrives? As you shove that tampon in or wearily stick a pad into your pants just before dragging your aching body to bed, do you really feel like living fearlessly? Or would it be closer to the truth that everyone around you should be living in fear as you deal with the next wave of cramps?

The phrases listed above are all lifted from recent ad

campaigns showing off the same old menstrual products which have barely changed since their inception.

Make no mistake. You are not about to read a lengthy diatribe against capitalism or these brands which, incidentally, I refuse to label 'feminine care brands', or – even worse – 'feminine hygiene companies'. Our periods do not make us weird dirty lady creatures. Don't you just *hate* those patronising aisles in supermarkets which have the label 'feminine hygiene' dangling above them in a halo of fluorescent lighting? Just call a pad a pad already! I also know people feel similarly about the euphemistic phrase sanitary towel – which implies bleeding women are unclean and of course it ain't no towel.

No, I truly am grateful to the Kimberley Clarks and Tambrands of this world, which, in the 1920s and 1930s, created sanitary pads and tampons for the mass market. Of course my life has been made immeasurably easier by not having to fashion my own rag out of scraps of material that I have to rewash all the time. I love tampons and sanitary towels as much as any woman can love those lady nappies that we interact with at the most painful time of the month.

But, taking a step back, doesn't it just make your mind boggle that before the 1930s, all women everywhere made their own period products? Millions of women across the world still do. But, with the advent of these companies, periods in the West became profitable almost overnight and the industry woke up to the fact that menstrual blood was big business.

Suddenly, men, as it was usually them (with the notable exception of Gertrude Schulte Tenderich, who, if you remember, bought the first modern applicator tampon from its male creator, Dr Earle Haas, patented it and began the company Tampax), became filthy rich from our 'monthly filth' – as they would infer in their sickly-sweet adverts. Not the early ones, of course, which used calm medical professionals to guide the new consumers through these new-fangled products and the process of buying period equipment.

The next period advertising trend came in the guise of sex education. A particular highlight being the 1946 Disney film on menstruation (look it up, it's a corker – you would genuinely think a period is the most soothingly lovely thing to ever befall a woman, who looks distinctly like Sleeping Beauty as the period dawns). And then comes the scene featuring graceful menstruating ladies in ballgowns. Of course that would be the natural next shot in such a bizarre movie.

Inevitably the advertising did change with the times. The 1960s and 70s brought in a whole new wave of feminism and the sanitary brands responded by promoting images of freedom created by their products to these newly emancipated women, who at the time were demanding all sorts of madness – like equal pay. Crazy.

But let me throw two dates at you: 1985 and 2017.

Apart from being the auspicious year of my birth, 1985 marked the first time the word 'period' was actually uttered on American TV, in a Tampax advert. Up until that point, sanitary brands hadn't been able to say the very word that describes their entire *raison d'etre*. Just take that in. Far more tasteful euphemisms, such as 'that time of the month', provided the language of blood up until that 'major moment' in period history. Shock horror, Tampax decided to call a period – wait for it – a period, to women who knew and had no problem with properly naming what was going on in their pants.

The honour fell to Courtney Cox (yes, the one from *Friends*, but at the time a relatively unknown actress). Her exact words within the advert were: 'Do you change for one week because of that time of the month? Still using pads? Then let me tell it to you straight. Tampax can change the way you feel about that time… Remember, there's a feeling with Tampax. It can actually change the way you feel about your period.'

The soon-to-be-superstar was pulling on a pair of deeply eighties leg warmers whilst wearing purple tights and a leotard as she said these words, while another woman, presumably also on her period, performed aerobics behind her.

Yep. Exactly what I feel like doing when bleeding: aerobics in the tightest clothing I can find. How *did* they know?

I contacted Courtney's representatives to see if the actress would reflect upon this moment thirty-four years on, in the hope that she would tell us a bit more about what went on behind the scenes of her history-making advert. Sadly the answer was an unequivocal no, despite my best efforts.

Fast forward to 2017, nearly a century after the first period products hit the mass market, and finally a period brand shows red liquid in an advert. The brand? Bodyform. The tagline? 'Blood Normal'. So far, so good. The internet erupted in praise as women took to their keypads to scream: 'Finally, blue liquid be done.' The period brand was hailed taboo-busting and revolutionary – all for showing a more realistic representation of blood. Real revolutionaries must feel a bit aggrieved – I'd imagine Emmeline Pankhurst might be rolling in her grave. Forgive my sarcasm but you see, before you get too excited, there's a catch...

The advert was online only. That's right, the revolution, it seems, was only deemed acceptable for the confines of the world wide web, and not any unsuspecting viewers watching their television sets.

I was curious. Did Bodyform do this because the media regulator in the UK didn't allow such content on television? This was certainly the assumption made by many who wrote about the limitations of the 'taboo-breaking' advert. So, I asked Bodyform. A spokesperson confirmed there were no plans to run the campaign on television, but denied it was due to sensitivity or the potential to cause offence. It's because the brand is now '100 per cent digital' and no longer runs campaigns on 'traditional TV'. To me, this seems like a missed opportunity to reach mainstream mass audiences who still gather around the box for key moments and could do with a dose of period reality.

Bodyform's own research, which their brand managers cited during the release of the so-called ground-breaking 'Blood Normal' ad, found that one in five women felt their confidence had been damaged because periods weren't discussed openly. Surely, there's an irony here?

Another queen in the industry, Always, have won awards for the 'feminist' messaging within their advertising, which tries to show how being 'like a girl' isn't an insult. The campaigns include montages showing teenage girls running, jumping and throwing like a girl. They're beautifully made, feel-good mini films – but you would have no idea they are about periods.

When I approached Always to ask why it had never depicted red blood in any of its TV advertisements, a spokesperson was able to avoid answering the question directly by saying: 'Currently, we don't show any product demos in our advertising, and therefore do not show either blue or red liquid.'

I pushed back and asked again if Always has ever shown red liquid in a TV advert and if not, would it in the future? Here's the reply in full:

> Although the brand may have included blue liquid in the past, over the last few years Always has refocused communications in line with their mission to empower women and build confidence. Because of this, product

> demos that would need red or blue liquid, do not feature in any adverts. The brand team has told us that advertising pads in the past has been heavily regulated, including a total ban to being covered under the 'decency and causing no harm' clause within ASA guidelines. This has obviously changed over the years. For the above reasons, we don't feel the brand is best placed to provide commentary for your book. Sorry we couldn't be of more use.

Surely, a major period brand is *perfectly* placed to provide commentary to a book about periods, but of course, it's entirely the company's choice to remain silent. However, when the folks who make their money from periods don't even show red liquid on mainstream TV, and are still pumping out adverts featuring girls running on the spot or being generally athletic while supposedly menstruating, it's no wonder that this conversation is still stuck in a rut.

In contrast, the Advertising Standards Association, the UK body in charge of regulating all ads and handling any complaints, was crystal clear in its response to me: 'We don't have any rules against the use of red blood in advertising sanitary products.'

So, the decision by any menstrual product brand not to show red liquid on TV is theirs and theirs alone. Frustrating, isn't it, that whilst the word period may now be used freely, that the biggest companies in the business – who as such, effectively control the narrative around menstruation – are still talking in hushed euphemisms and secrecy.

Yet, whilst these major brands are unwilling to represent periods in their glorious bloody truth in their advertising campaigns, they are at least piggy-backing onto the activism of feminists in this space. Always, for instance, at the time of writing has donated over 14 million pads to UK schoolgirls in need, donating one product for every pack purchased. Tampax has publicly committed to donate a further five million products. Bodyform has promised in a three-year-plan to donate 200,000 packs of sanitary products by 2020 to help women and girls access free period protection.

However, we mustn't forget the horribly ironic truth that if periods had never become a money-making business in the first place, and instead governments issued pads for free (or at least ensured they were as readily available as toilet paper) *there would be no period poverty*.

Ever since their inception, the giant brands (which make good dollar from your blood) have shaped the international representation of periods. Blue liquid instead of red; women eating creamy clean yoghurt; crisp white clothing rather than stained pyjamas and hot water bottles; cheery, inspirational scripts instead of grimacing, pain-killer fuelled monologues. The reality just isn't seen as desirable in the brands' mission to flog as many units of absorbent cotton wedges as possible.

It perpetuates the age-old myths, just dressed up in modern clothing in fashionable 'empowering' language. Yes, we might have moved on from these brands telling us to hide our smelly periods (note one of Tampax's first adverts, promising to eliminate 'odour and embarrassment' whilst permitting 'daintiness at all times' – *lovely*), but there's never been a true-to-life representation of how women really look, feel and act during their periods. Yet we just absorb this weird, flowery fantasy of mother nature making her wonderful monthly visit without much thought.

Art historian Camilla Rostvik, currently writing a PhD snazzily entitled 'The Painters Are In' on the history of menstruation art since 1950 (shout out to the Leverhulme Trust for funding her fascinating-sounding work), believes that the power of these menstrual product brands is absolute. Here's what she has to say:

> Where do normal women get their messages about periods from? It's adverts. I wish people would get their messages about periods from the activists who are breaking new ground in this area and trying to help people, rather than those selling their products. Activists who are trying to end period poverty or make people feel positive about their periods are the ones holding these big companies to account and never make women feel shameful about their bodies and menstruation.

This is an industry that can even turn a public health disaster into a money-making chance. Following a spate of toxic shock syndrome incidences in 1980s America, when Procter & Gamble had to recall its new, highly absorbent tampon 'Rely', after concerns that a chemical used in production had caused a spike in cases, the big brands spun this into a fantastic commercial opportunity, as Camilla explains: 'Suddenly the other period brands realised women would need and want lots of tampons instead of just one super absorbent one. [which had been the trend at the time of Rely]. The heavier the period the better – more blood equals more tampons which equals more money spent.'

The recall of Rely meant a move by the major brands towards telling women to regularly change their tampon for health and cleanliness reasons, but it also meant women would need more tampons, which conveniently equalled more money in the bank for these businesses.

But Camilla is also at pains to stress that some of the brightest, most feminist women work on these accounts in advertising agencies – but they can only push the message so far. Of course male advertising executives don't compete to create adverts for periods. *Doh*.

It made me think about who else influences large swathes of society, especially in the age of social media: celebrities. Alongside brands, they hold a huge sway over the population, but, and here's the weird thing, if you cast your mind back can you pick out a major celebrity who starred in a super cool period ad? Nope. Me neither. And unfortunately Courtney Cox doesn't count, because when she first mentioned 'period' on national television, her star was yet to rise. Perhaps, if she was already famous, she might not have agreed to do the ad at all…

But if she had shunned the ad, she wouldn't be alone

in side-stepping menstrual products as desirable brands to endorse. In fact, Karen Houppert, the American journalist and author, made a surprising discovery in her book about periods, *The Curse*, released in 2000: at the height of their fame in 1997, the Spice Girls reportedly turned down Johnson & Johnson's offer to appear in an advert for a new sanitary pad. Karen's source, an insider who wished to remain anonymous and worked at the advertising agency McCann-Erickson which handled the launch, said of the Spice Girls' refusal: 'Even this group, all about girl power, wanted nothing to do with pads.'

You will remember that the Spice Girls was a pop group who happily appeared in an ad for Walkers crisps, so they certainly weren't against endorsing brands. But periods? No, this was seemingly one step too far – even for the band that became a symbol for girl power. Perhaps it was a purely commercial decision. Perhaps the Walkers' cheque book was a lot bigger. I approached former Spice Girl, fellow endometriosis sufferer and all-round lovely woman, Emma Bunton, to tell her recollection of the story. Alas, she too declined the opportunity.

According to Karen's advertising insider, many athletes, film and TV stars were approached for Johnson & Johnson's new sanitary pad launch, but most refused to be the face of this new product. Are you spotting a trend? (Sorry, I'll try and ensure that will be the last punny use of the word spotting.) The collective silence and squeamishness about periods across society, a behaviour perpetuated by the major brands, has created such a secretive vibe around the truth about menstruation for so long, that very few high-profile folk will be associated with any part of it.

UNMENTIONABLE MENSTRUATION

Normally, older brands are kicked into shape by new contenders. In most industry sectors, innovation gives rise to better consumer choice and the best possible products. The smartphone practically killed the digital camera. Spotify and iTunes practically killed the CD. Specialist tights brand, Heist, is taking on the hosiery giants. I could go on, but the period sector ain't like that. Sanitary products remain unchanged and the brands reign supreme.

Let us not forget those truly awful tampon bins in most public toilets around the country. Nope, they haven't improved or changed with the times either. No offence to those who manufacture them, but even the ones with hand sensors still break, are annoying to use, cumbersome and totally and utterly gross (yes, there's usually a stray sani pad sticking out the top). Don't even get me started on the people who build toilet cubicles so narrow that you have wiggle around the swinging door and brush your thigh against aforementioned gross sanitary bin as you attempt to enter. It really is the pits.

Baroness Martha Lane Fox, the investor and entrepreneur best known for creating and selling lastminute.com (and whose words head up this chapter), isn't surprised by the lack of innovation in such an expressly female space:

> It comes back to who is controlling the financing and funding of ideas in our society. The most common people who seek me out are women who have worked in the care space or public sector, have an idea but cannot get any traction with the VCs [venture capital firms who invest in new companies]. Only 10 per cent of those who work in VCs are women. And far fewer are the partners in those companies who really hold sway over what ideas they invest in.

Consequently, Martha believes there could be many ideas out there in the period space, but they aren't cutting through to the mainstream – not because they don't exist, but because they can't secure funding. For her, the solution to filling in many of society's blind spots lies in getting more women and people of diverse backgrounds into those decision making positions within VCs and other organisations that fund new ideas – even the unsexy ones.

Preach, sister.

You might recall how a cool young British company 'Sanitary Owl', now rebranded to 'Dame', reduced the famous investors of Dragon's Den to giggling schoolchildren in 2017. I would go as far to say that notoriously stony-faced Peter Jones looked as if he was about to be sick. Co-founders Alec Mills and Celia Pool appeared on the programme to try to secure funding for their period subscription box service, which included all types of sanitary ware from Mooncups to tampons. I must say, it went wrong from the off with the voiceover script saying, 'It may seem counter-intuitive to launch a product half the population may never have use for.'

Really?

The last time I checked, 32 million women is a pretty large market. After that, it only got worse…

As soon as tampons were mentioned, Peter Jones said he felt 'very uncomfortable'. The so-called dragons became weirdly hysterical, embarrassed and giddy at the 'odd pitch'. People at home took to Twitter to denounce the bad period vibes and to air their complaints about the reactions.

Needless to say, Alec and Celia walked away with nada. Not a dime. And this was the first time a period product had been pitched in the den. As Alec described the scene soon after, the BBC and the sniggering dragons had squandered a chance to normalise something so terribly natural. He told Laura Coryton, the tampon tax campaigner, that he was upset about the way in which the show was edited and that the producers focused on 'funny and squirmy bits' of the pitch, cutting out the entrepreneurs' business strategy and the discussion of their aim to help homeless women.

Alec said: 'It was a real shame for the 32 million women in the UK who could probably have benefited from a little more period-positivity.'

It's worth noting that, at the time of writing, Dame is

still going, having secured funding elsewhere, pivoting its business model completely to a genuinely innovative sounding product: a reusable tampon applicator.

Now, I haven't tried it, but it's so bloody logical when you think about the crazy amount of plastic waste created by binning an applicator every time we use a tampon. That's the kind of innovation I am talking about. Why has no one else come up with something similar until now? Recently, a spate of new products have sprung up as companies are finally cottoning on (couldn't help it) to the business potential of reinventing women's experiences of periods for the twenty-first century.

In 2014, the American brand Thinx launched their period pants. The super-absorbent knickers allow women to free bleed directly into a specially-designed gusset (side note: is there a better word than 'gusset' in the English language? Perhaps 'flange' but you get the gist…). Rival period pant brand Modibodi debuted the year before in Australia and now WUKA (Wake Up Kick Ass) is giving both ultra absorbent pant brands a run for their money.

To date, Thinx has flogged approximately half a million pairs of knickers around the world, which at roughly £30 a pop isn't a small sum. Yes, they do require washing at a time of the month you generally cannot be arsed, but they still represent some rare innovation in the period product space. Plus, the recent launch is definitely worth reflecting on because of the shitshow that occurred…

When Thinx submitted its adverts to Outfront Media, the company in charge of flogging the New York subway's advertising space, the responses were almost comic. The first problem? The fact that the brand was using the word 'period' straight up in its promotional material. The advertising agent from Outfront Media objected to Thinx's slogan, 'Underwear for Women with Periods' saying they couldn't run the copy as it currently read. An email to Thinx read as follows: 'Is there another way for you to position this? Perhaps there is a creative way to say it.' Um. Like what?

Secondly, there was an issue with the word 'fluid' being used in the blurb. Again, the agent from Outfront Media took issue with it and this time asked one of Thinx's staff (this time by phone), 'What if a kid saw the ad?'

Reportedly, he also said that the pictures of the grapefruit – meant to symbolise the vagina and eggs in place of a woman's ovum – were too suggestive, alongside the female models in underwear.

Right. This is the same agency which has accepted ads featuring women in strappy bikinis to advertise fitness products, boob jobs and swimwear! The women in Thinx's ads were wearing polo neck jumpers with their full knickers. Not a tit or vagina in sight.

If anything, the adverts, in light of my point about euphemism and disguise, were still too tame. And yet still they managed to provoke the same sort of objections as if periods deserved the same treatment as dirty porn. Thinx's director of marketing, Veronica Del Rosario, recalls the agent slamming the phone down on her when she questioned his problems with the ads saying: 'Don't make this a women's rights thing.'

Finally, in 2015, we had a brand-new entrant into the period space. But this debacle still happened – and the madness continues.

You would hope that, away from the period product brands, other cutting-edge companies and organisations wouldn't have such a period blind spot. But I'm terribly sorry to inform you, you would be wrong...

Let's take a look at Apple, for example. Currently one of the richest companies in the world, with more money in the bank than most governments and more data at its fingertips than most of us will ever know – and more than is probably wise for a private company. You would hope that such a beacon of creativity had enough insight to, you know, not forget periods exist.

Yet, when Apple released its health kit app in 2014, marketed as a one-stop shop for people to track their vital statistics, you would hope that something that affected half the world's population might have its own section. But no, not even a footnote.

Apple forgot about periods. Simple as. Oopsy!

But how did such a major oversight occur? Martha Lane Fox, who incidentally now sits on the board of Twitter, has the answer:

> Do you think Apple would have released its much anticipated 'Health' kit without the ability to track periods if there'd been a woman high-up in the organisation? I don't. And is this why the big tech companies haven't addressed the issues that are predominantly faced by women on social media? Trolling, harassment, death threats? If there were more women at senior levels in these companies, perhaps problems would have been solved sooner.

Since then, the embarrassing error was corrected and women can now happily track their monthly cycles, feeding even more personal data to the world's most successful company who temporarily forgot about them. Happy days.

Let's talk about another tech company: Fitbit.

Fitbit was founded in 2007 in San Francisco. The company sells a range of wearable products which allow people to track their personal metrics with a view to measuring their fitness: number of steps, quality of sleep, heart rate, etc. You get the gist, perhaps you're even a customer.

However, it took until 2018 – more than a full decade after their inception – for the company to breathlessly announce the launch of its 'female health tracking' system in an excitable blog post, proclaiming: 'One of your most requested features is here!'

Now, I accept that period and ovulation tracking is not a key fitness metric – but if it really was one of the company's most requested features, why did it take a full eleven years to make its debut? *Chin-stroke time.* One can only hypothesize that perhaps there were no women at the helm of the product development and design.

However, the launch of Fitbit's female health tracking system soon blew up in its face as women reacted around the world to the fact that the system limited a period to 10 days or fewer. Women with longer periods were prohibited from logging them. A spokesperson for the company confirmed to the BBC that 'currently a period must be less than eleven days'.

And instead of just immediately fixing this arbitrary limit, Fitbit, in its infinite wisdom, asked those 'concerned' to comment and vote on its suggestions board.

As you can see, even away from how the menstrual companies shape the global conversation about menstruation, periods are either forgotten about by brands who should know better, or treated as a poorly understood afterthought.

THE PERIOD EMOJI

I feel compelled to point out to you one other area of modern life where periods have been overlooked once again: the emoji keyboard.

At the time of writing, there is no period emoji. There's a poo emoji. There's even a relatively new breastfeeding emoji. There are twelve different types of train emoji. But periods? Absolutely not. Take that in for a minute.

Doesn't it strike you as odd? Instead, we are forced to improvise, often using the Japanese flag to denote a blood drop on a white patch. Sorry, Japan.

But, *breaking news* at the time of writing, there should be a period emoji of sorts by the time this book is published, after the extensive and sleuth-like efforts of the girls' rights charity Plan UK and the NHS.

Yet, until now, in the strange new alphabet of this modern visual, digital world, the period emoji has been missing. If you aren't partial to emoji use, then you might be rolling your eyes, wondering what the big deal is. But for millions of people, emojis are key to their daily communication – there's a palm tree, an octopus and a spanner, but women and girls don't have a symbol to denote something that happens month in, month out.

Carmen Barlow, the digital strategy and development manager at Plan UK, decided this was nonsense and to take action – she began her mission in November 2016:

> We thought it was very strange that there was no period emoji. It's weird not to have an icon for something 50 per cent of the population has 25 per cent of the time. It felt strange to have a lack of representation in this way, especially when we know the taboo and stigma that stops women from talking about their periods. My team and I were brainstorming the hard-hitting issues we were preparing to tackle when we started thinking about a more light-hearted way to address period stigma. How we could create a less serious and heavy conversation about period taboos to reach a wider audience.

It set Carmen down a long, odd road. The road to the Unicode Consortium. A group I have never heard of and I'm pretty sure you haven't either. Essentially, these are the overlords of your emoji keyboard. Based in California, this board – which is a not-for-profit organisation – consists of individual members (mainly men) from small and large tech companies including the likes of Facebook, Google and Apple, that are tasked with developing and maintaining a common standard of text in all modern software products. And this includes emojis, so that there is a common keyboard around the world. They decide what's in and out. And there are proper application processes.

Carmen's team set about putting a few period emoji designs out there online to see which won the popular vote. The majority, supported by more than 55,000 people, depicted a neat-looking pair of white knickers with a single red droplet. The design drew international interest and, feeling buoyed by the online support, they put it in for consideration to the Unicode crew at the start of 2017. It was rejected.

The reason? Not universal enough. Carmen was understandably dejected and annoyed. 'How could a white pair of knickers with a drop of blood not be universal enough?' she asks. 'Everyone around the world would know what that image was referring to. This was never said in the formal rejection, but I couldn't help feeling it was actually turned down because of a squeamishness to do with blood and periods.'

I contacted the Unicode Consortium to ask them for

more details about this rejection and why it had taken so long for a period emoji to appear. I am still waiting for a reply.

But Carmen's quest was not over. A woman from the board got in touch and quietly recommended that Plan UK join forces with NHS Blood. A partnership was swiftly formed and another bid went in, this time with a single red droplet of blood as the design.

It passed.

And while the application clearly states that it was accepted as both a marker for blood and menstruation, when it launches this year it will simply be known as the blood-drop emoji rather than an icon expressly representing periods. Again, make of that what you will. Heaven forfend periods actually had their own emoji. No, they must be diluted and share the stage with all blood.

A pragmatic Carmen is pleased, calling it 'a big step in the right direction' but says there is still more to achieve. 'I believe the consortium didn't want an explicit menstruation emoji but I am hoping that will change. The next step has to be a tampon or pad emoji. The breastfeeding emoji is quite new so I remain hopeful of more change to come.'

So, watch this emoji space.

Personally, I'm not holding my breath for a mini tampon to find its way onto my screen anytime soon . . .

When all these moments in period history come together, you can see why regular women still don't feel comfortable talking about their menstrual blood. Why they think periods are dirty things which should be hidden or, better still, forgotten. And why innovators don't even bother coming up with new period solutions, for fear they will be laughed out of the room. When the big brands tasked with talking about periods still use euphemisms, or worse, don't even mention periods at all in their ads, it's clear that unless we do something about it, this taboo is here to stay.

Martha Lane Fox said something to me that I haven't been able to stop thinking about: if people had genuinely continued to innovate around periods, we would be in a post-pain era. Women would have devices or drugs that would ensure that their periods gave them as little pain as possible.

But again, the funding still isn't there for medical research in this area, which until recently was dominated by male researchers and focused on problems that (most of the time), men experienced too.

It has been widely quipped that if men got pregnant there would be an abortion clinic on every corner. There is a similar truth about periods. Just think of the advances in products, medicine and the macho 'man-pon' style adverts that would ensue, if men had periods too.

A CALL TO ARMS

We women need to have our wits about us. I'm pointing out all these things because it's important to remember who is still controlling the narrative when it comes to periods – not because I want you to ditch these brands and mould your own Mooncup at pottery class. And crucially, we need to remember what *isn't* being said by the big giants of the sanitary world.

It matters that Apple forgot about periods.

It matters that Fitbit thought periods could only last 10 days.

It matters that a period emoji was turned down.

It matters that a leading investor giggled like a child when a period brand came to pitch to him.

It matters that the New York subway didn't want a new period brand to use the word period in adverts.

It matters that leading older period brands still haven't shown red liquid in TV ads.

Perhaps you could dream up your own, very real, period ad… Mine would mainly involve grease: greasy hair, greasy Chinese and a greasy face. Oh, and many profanities nestled within a sassy monologue about my period vibe. And low slung tights.

Then compare and contrast your ad to the ones made by your preferred tampon or pad provider and have a big belly laugh at the differences. Rejoice in the reality you have mentally created and let *that* be the narrative that dictates how you think of periods.

Even better, go one step further and film your real period advert. The rougher the cut, the better. Someone needs to start challenging the big period players.

Go forth armed with the knowledge of what isn't happening in the period space and what should be – the language and icons which are used, but also those that are not. Question why.

Hopefully, the insights of this chapter might provoke a wry laugh next time you're in a supermarket's 'feminine hygiene' aisle.

And now you know the truth behind the battle to secure the period emoji and this small but significant step for womankind – really relish using it each time you do.

CHAPTER TEN

SEX BLOOD

'If there is a man who lies with a menstruous woman and uncovers her nakedness... both of them shall be cut off from among their people.'

Leviticus 20:18

We have come this far without talking about it but now's the time.

Periods and sex. Otherwise known as period sex.

Every weekend where I live in London, there's a determined dude selling posters near the tube station. It's a distinctly retro situation, with many of the posters featuring nineties bands and nineties lads' banter. Recently, one gem caught my eye. It featured a hot brunette woman, looking coquettishly over the edge of her sunglasses, wearing a tight skimpy black dress, and crouching down with one leg casually open to the side – like you do. Above her head it read: '15 Reasons Why a Beer is Better Than a Woman.' Naturally intrigued, I read on.

The fifteen reasons snaking down the side of the woman's body included: 'A beer always goes down easy' and 'Beer is always wet'. The charming list culminated in this final point: 'You can enjoy a beer all month long.'

Ladies, it seems we haven't moved on from the Biblical teachings of Leviticus after all, in which the good book states: 'If there is a man who lies with a menstruous woman and uncovers her nakedness... both of them shall be cut off from among their people.'

Breaking news: you can have sex with a woman all month long. And that includes when she's menstruating. You will not be cast out of your family and friends' tribe. (That will only happen if you are found in un-ironic possession of this poster.)

Most people will still not publicly admit to period sex – and especially to liking it. Period sex is still widely viewed as a pervert's paradise, a fetish and a weirdo's delight. Often, the focus is only on the man's point of view and how *he* feels about it.

Forget about us females, and how we're feeling. Whenever this uber-taboo topic arises, women are still placed in the passive role – with the assumption that we should simply be quietly grateful for any man brave enough to let his manhood be sullied with our monthly bleed.

While period sex is definitely a fetish for some, it is also incredibly normal for others. But for a whole other group of people, it remains a problem about which to fret.

So, what is the deal with period sex? How should we think about it? What are the facts? What's fiction? Where should you stand? And have you, perhaps, been thinking about this whole issue in the wrong way all along?

A brilliant range of women have shared their views and vivid experiences with me for this chapter but, unsurprisingly, they have asked to remain anonymous. However, rest assured that when it comes to period sex quandaries, you are not alone. Believe me. So let's ride it out together, eh? Sorry. You know I can't resist.

Petra Boynton is an agony aunt and has been for many years. She receives several questions a month about period sex from all ages. Every. Single. Month. 'Is sex on your period safe?' 'I want to have sex when I am menstruating but I'm worried my partner will think it's icky.' 'Can you get pregnant during period sex?' 'I don't feel like sex during my period but my partner wants to.' You get the gist, there's continuing angst about what the deal is when periods collide with intercourse. Or don't.

More recently, Petra has received questions about anal sex being a substitute for vaginal sex during menstruation. This genre of questions brings to mind the response from my friend Keira when I asked if she would ever have period sex:

> No. Never. Wouldn't. Couldn't. Though at school there was a very horny girl called Stacey and one night we went down to Brighton for a night out and she was badgering this bloke we knew, Tom, for sex. She wouldn't let it go

> and at one point she said, 'I'm on my period but if that bothers you we can go up the bum.' Nice. He didn't cave and I was quite relieved as we were all sharing a room in a hostel.

Stacey may have been 'very horny' but she was also trying to do what many women feel they ought to during their time of the month: pay the period sex debt.

As another friend of mine put it: 'I never have sex on my period. Definitely not… no need. Why? The mess! Surely that's what blow jobs are for?!'

Ah, the consolation blow job. Or the overly generous offer of anal sex in lieu of normal service. They are both part of a pretty weird phenomenon: women, at often their most uncomfortable time of the month, offering to pleasure men to make up for their periods being in town. And getting nothing in return.

Now, of course if a woman genuinely feels like doing those things then all power to her. If she loathes period sex because she doesn't feel like it or hates any mess, but still fancies giving her partner some action, then again, all power to her.

But if she's fellating them out of some sort of guilt or compensation – or even worse, because she thinks her partner would be repulsed to have vaginal sex with her when bleeding and she's actually horny, then that's messed up.

By and large women tend to fall into three categories, which I will describe below and explore in more detail…

1. The woman who never wants period sex

She never desires red sex, end of. She feels least sexy during her period, and it's understandable – she's in pain, she's bloated, she has the runs. In short, there's a lot going on down there which isn't conducive to wanting to rip her knickers off and get intimate. She, and I include myself in this description, need to retreat to their own modern day red tent – not out of shame, out of preference.

It's totally normal. And for many women, periods offer a welcome respite from any sexual expectation and other physical activities. They are a chance to just *be*. A newer girlfriend in my life, Jane, confides that she feels at her most vulnerable during her period and couldn't consider doing it during her heavy days. As she puts it:

> My husband is totally chilled and unfazed by menstruation (he is not really squeamish about anything) and I reckon he would be up for sex the entire duration of my period! But for me it has to be at the beginning or end of my period, mostly for reasons of comfort but also because I feel it's a private and slightly vulnerable time of the month when I am heavily bleeding – I feel I wouldn't want to share the experience too openly with anyone. It is MY period, not anyone else's.

In this group of women are also those who hate mess. Loathe it. And in a life filled with mess, often courtesy of children, they just can't handle cleaning up anymore.

Cue the experience of another pal, who has four children and cannot abide any more housework, who told me this tale:

> Yep, I've had period sex. In the early days of having young children, you take your chances when you can – and sometimes needs must. In fact, we did it again quite recently after not needing to for a long time – and then remembered why we don't. The bloody mess. How can two people create that much fluid? He looked like he'd survived some sort of medieval battle with his (figurative) reddened dripping sword. I looked like I'd been part of some ritual initiation ceremony. The bed. The sheet. The mattress. We agreed no orgasm is worth that washing. There's a reason people don't do it.

Sometimes, this group of women will give out sexual favours if they feel like it during their period of abstinence – but by and large, they are happy with their position of 'no period sex' and their partners have received the memo loud and clear.

2. The woman who doesn't know she wants it

She doesn't think that she wants to have period sex, when actually, she might. She's been so infected with the wrongful shame surrounding periods that she can't tune into her body's needs during menstruation.

So, I'd like you to meet my friend Edie, who displays a degree of ambiguity when it comes to period sex.

Sometimes, she actually feels horny on her period, but would never dare allow herself to explore those feelings as she worries her boyfriend will 'find it gross'.

> I haven't ever had sex on my period and wouldn't want to personally. I would feel very self-conscious about mess on the sheets, on him (ahem) and would worry about him finding it gross, even though he says he wouldn't. I also don't feel very sexy when I'm on my period (bloated, bit uncomfortable, not that sparkling clean), so feel like I wouldn't want to start feeling all sexy then anyway...

How many women, if they stop and properly think about it, are actually in the same boat? They can't even begin to think about broaching the conversation with their other halves out of paralysing fear of how disgusted they will be. And maybe they *will* be disgusted, but that doesn't mean you can't at least discuss your sexual needs and see if you can work through the problem. Any decent partner would loathe the idea of their loved one denying themselves sexually at their expense, without even the chance to rectify the matter.

3. The woman who knows she wants it and has it!

Yep, she has no qualms about some blood on the bed sheets, she knows what she wants and no man, and no menstrual fluid, will stand in her way to enjoying period sex. This woman might just demand it as her right, or she might actually crave it because it feels damn good – or both.

Speaking of 'decent partners', one of my pals, Clara, actively uses period sex as a test of the calibre of a potential male partner. She falls firmly into the period sex camp, seeing it as her right. She isn't remotely political or feminist, more a pragmatist through and through: 'I find a man's ability to deal with a bit of blood is an important test. The assholes are never relaxed about you accidentally coming on or being on during sex and make it into a huge deal.'

And as a different girlfriend of mine (who is as equally pragmatic but also very passionate), puts it:

> YES! Of course I have period sex. Absolutely! That is one quarter of your sex life you're missing out on if you're not... For me, the idea that you shouldn't because you are emitting liquid, especially in a sex act where a man literally ejects liquid, is so much about the shaming of women's bodies.

It's pretty sobering, when you think about it – that women the world over, horny during their periods but lying low, are denying themselves a quarter of their sex life. Especially when it's over a small amount of red liquid in an exercise that's all about bodily fluids.

And in fact, there is medical evidence to support the idea that period sex is actually good for a woman. When a woman orgasms we release endorphins, which are the body's natural painkillers, so period sex can help ease period cramps. Who'da thunk it? (The most romantic story on this little nugget of information is yet to come, just you wait...)

Period blood can also act as a natural lubricant, making sex more pleasurable and easy all round (that is, if you have remembered to lay out that trusty absorbent old towel on your bed first, facilitating the clear-up job afterwards).

And some women really do report feeling their horniest mid-flow or just as they are about to come on their periods. What is hotter than having sex with someone when they are at their horniest?

The positive sides of period sex are so rarely discussed that I doubt most people, men or women, have a clue. Earlier in the book you met Jillian Welsh, the Canadian performer who ended up nearly doing jail time for stealing her partner's saturated bed sheets.

A fellow writer mate of mine ended up in a similar bloodied scene but instead of fearing it, she laughed it off in true British style, and it proved to be the basis of a very important relationship:

> After I separated from my ex-husband, I rented a very basic studio flat on my own. I had started seeing a new man, but it was casual. We were so excited about my flat and the unfettered nookie that it would allow, that he came over the evening I moved in. I'd only got the keys a couple of hours earlier and hadn't unpacked anything, but I'd managed to make the bed, with some new white

> bedding. The man and I had lots of celebratory drinks and then got down to it. Unfortunately, mid-coitus I rather generously started my period. By then it was late at night, we were very drunk and the lights were off. We stumbled around and tried to clean ourselves. 'This bath drains slowly,' he slurred as he showered off.
>
> Soon we passed out and awoke the next day to a scene from CSI. There were violently blood-stained sheets, a bloody trail to the bathroom, bloody footprints, and the bath (blocked, as it turns out) was half full of deep red water. It was absolutely disgusting and embarrassing. My casual lover was quite put off, but only for a while. Three years later we married.

Period sex can be excellent for the woman, both in terms of sating natural desires and also, helping her physically. It can also bring partners closer together, as there are no times of the month they are out of bounds to one another. And the woman feels no shame about her body in her own home. She can even feel emboldened by such a lack of squeamishness.

What I am keen to stress, regardless of which group you, or the women you know, fall into, is that period sex is a real choice and women should feel totally free to do whatever they want, as opposed to paralysed by shame or fear.

But it does take (at least) two to make period sex happen.

It's time to talk about the men in all of this, often our

willing or unwilling partners in this act. They also fall into a few categories of their own:

Those who actively seek period sex and are aroused by it.

Those who do it for their partners.

Those who boast about doing it.

Those who don't give a shit.

Those who just don't.

Let's take 'em in turn…

1. The Bloodhounds

Yes, these are the men who actively seek out period sex. In writing this chapter I am privileged to have made contact with the American journalist Maureen O'Connor, who writes for *New York Magazine*. For it is she who, in one of the only high profile pieces about menstrual sex, coined the term 'bloodhound'.

Isn't it just exquisite?

Catching up by phone she tells me: 'I made it up on the spot. I said to my editor while I was researching the subject, we have to call this piece "Bloodhounds". Of course, I spoke to lots of women who were into period sex, which, as a woman who loathes sex on my period, surprised me. But I needed to name the men who were really into it.'

Her fascinating article ended up being called: 'Bloodhounds: They're obsessed with period sex.'

Maureen cast her net out and it wasn't long before she heard about several guys, including one male pop star, who was into period sex in a big way – an all over his face kind of way.

And then there was the guy who only wanted period sex with his girlfriend, specifically requesting their sexual meetings to be timed around her cycle. The same man has also ruined a classic dessert for Maureen forevermore. You see, he named it 'strawberries and cream' sex. Delish. As Maureen puts it:

> The food references make it very visceral. For me personally, taste is a sense I turn off during sex. I usually try to ignore the taste of sex. But the idea of some people relishing it, especially period sex, is so different to me and, if anything, makes me feel more comfortable about my choices. We are all so different. But I will always think of strawberries and cream in a different light since writing that piece.

You are not alone. Wimbledon is off from now on.

The guys she spoke to who expressed a fetish for period sex all expressed a desire for getting messy during sex. That was the turn on. Some may also get their thrill from the idea of dominating a woman at a potentially vulnerable time. Maureen, in her enviably wide writing portfolio, has also written for *New York Magazine* about another sexual taboo: rim jobs. The idea of men feeling dominant by going down on a menstruating women brings to mind an idea she heard discussed when researching all things 'butt'. (The headline on this gem was: 'Warning: A Column on Butt Stuff'). 'People can view anything as domination if they want. But only straight white men could convince themselves that they are dominating a woman when they are literally eating her shit. Who is really dominating who?'

It's a fair point.

Agony Aunt Petra says that it is a relatively small group of men who have sex with women during menstruation because it's a fetish. And even then, it's all about the blood, not the period:

> For those with a fetish or kink it isn't necessarily about period sex, but may also be about bleeding more generally – especially things like virginity loss – there's a genre of porn for that with much fake blood and even more faked pain reactions. Some people do like the look of blood or the mix of blood and semen. Or to see blood on their bodies or clothing.

It's at this point I have to confess, I've let you down. Badly. Up till now, I haven't flinched away from reading or watching anything during my research for this book. But, as much as I love you, and the idea of smashing the period taboo, I couldn't bring myself to sully my internet history with a terrifying search for period or blood porn. I hope you don't feel too cheated and we can still be mates.

Moving swiftly on…

2. Bleeding Love

I kid you not, there are a small group of men out there having period sex for the good of their women. In two kinds of ways.

As previously mentioned, the endorphins released during orgasm can help alleviate period pain. Maureen actually met a legendary man who isn't particularly into red sex but does it to help his wife's cramps. This is a twenty-first century romcom waiting to happen. Maureen says, 'He told me he looks at it like he's providing a service to her at that time of the month. It was very strange but also very sweet. And actually, kind of romantic.'

Chivalry just got a whole load more interesting.

And there's the menfolk who aren't particularly into period sex but their women are. Big time. So much so they feel they can't let them down and they end up getting turned on by how horny their partners are. Maureen met a man in this camp who managed to change her relationship with another favourite food of hers:

> One blood-averse man described women who got very horny during their periods, and their arousal aroused him. The ability to trigger cascading orgasmic freak-outs, he said, was incentive enough to perform cunnilingus on vaginas that tasted 'like very rare steak' and post-coital imagery he likened to 'human carnage'.

Needless to say, Maureen, who is a major fan of very rare steak, hasn't eaten it in quite the same way since.

3. Blood Braggers

Now these are the men who see themselves as sexual connoisseurs, adventurers, experts if you will. And what eating offal is to committed carnivores, eating period blood is to them. Lovely. They see themselves as being so comfortable with the female body, that being aroused by period sex is a clear sign of their sexual prowess. Red sex isn't a fetish per se, more a display of their vast sexual experience. It's like they have ascended to the highest levels of sexual capability and are a sex master now they have overcome the red stuff.

But there is also a curious subtext to these guys' comfort with bloodied bonking: it's a sign of their masculinity.

And it's not just men who feel menstrual sex is a barometer of masculinity. Some women told Maureen that 'a real man goes down on you on your period and they love it'.

How a period ends up being a barometer of a man's manhood is totally extraordinary – and yet somehow, in some men's minds, it does.

4. Bleeding Agnostics

Let's give it up for the men and women who are non-plussed. These are the men and women who don't really mind if they need to lay a towel down. If their lover is up for it, and there happens to be a period in play, they don't care either way. These men aren't into showboating about period sex. Nor does the blood do anything for them, except perhaps add a touch of welcome lubricant. It's just sex at a slightly messier time in a woman's cycle. This means they haven't been indoctrinated by religion or wider society to see women as shameful dirty animals while menstruating, best avoided until cleansed. Nor do they think of them as gross or kinky for wanting sex during their flow.

In turn, this relaxed attitude liberates the woman, allowing her to feel like she is always attractive and doesn't have to feel shame about her body for a week or so every single month. And so period sex loses its weird X factor, in a good way. Whisper it, it can become a non-event. With some giggles along the way.

Of course, a bleeding woman would still have to feel up for it herself, which in some people's cases will never happen. But partnered with nonplussed lovers like these guys, at least they will know for sure that they are not having period sex for the right reasons, rather than stopping themselves out of fear of judgement and shame.

5. No Bloody Way

Under no circumstances will these men ever touch a bleeding vagina. Ever. With a hand, a willy or a tongue. Never going to happen. These are the guys who laughed and spread stories about men who have gone there and lived to tell the tale.

They are absolutely appalled at the idea of period sex. But I do have some sympathy for some of them.

Screw the guys who genuinely believe women are

impure and filthy hussies during their period because of something a rabbi or imam said. Or something a biology teacher never said and should have done. I am not sparing a thought for those men who think we should skulk off and segregate ourselves until we are clean enough to be worthy of their touch. (Except when we are offering them one-way sexual favours. Convenient, much eh?)

But I do have time for those men and women who are just scared of blood. All blood. Not just the uterus-infused type. Blood, like urine and faeces, is a bodily waste product after all. We are trained to wrinkle our noses at it and abhor the sight of the red, yellow and brown stuff. Blood is also terrifying. It is literally the stuff of horror movies and death. Red signifies danger. You only normally see it in times of crisis and so it's natural that some folk cannot shift those deeply held and perfectly natural associations.

The few women generous enough to share their period sex stories in this book, even if they enjoyed said sex, do also liken the aftermath to a murder scene. They just manage to laugh about it. Whereas for others, blood up the walls would be the biggest turn off ever.

Sarah, another more squeamish friend, shared this charming teenage memory:

> I am not a fan of period sex but my first boyfriend was (did I ever tell you about him? Mad. I think he was watching hardcore porn long before mobile phones landed in kids' backpacks. He asked me to sleep with his brother

> whilst he watched. HE WAS 17!!!! And no, I didn't). Anyway, he had a beard (maybe he wasn't 17 after all now I've just written that sentence) and used to like going down on me when I was 'on'. He used to come up for air looking like Blue Beard. Rank. We tried a few times but not my thing.

And that's OK. It is never going to be some people's thing – men and women. And it can change over time too. One of my closest girlfriends used to have period sex all the time, but since she hit her late thirties has developed a serious aversion to all kinds of bodily fluids. Even her own. She can't bear to lie in her own juices post-sexual activity, never mind anyone else's. Period sex to her now would be a total anathema.

But there is a distinction to be drawn between those men and women who think women are disgusting because of their period and those who think blood is disgusting full stop.

It's hard for women not to interpret a partner's aversion to blood if they are open to period sex, as an aversion to her – thus beginning the whole cycle of shame again. That's why I am raising this distinction, so that you can consider the real reason behind your own possible aversion. And if it comes across wrong to the woman in your life, you can also rectify that and make her feel beautiful during her period – not disgusting.

In an attempt to move the conversation on about period sex and normalise it, in 2018, Thinx, the absorbing period knicker brand I wrote about earlier in the book, launched a Period Sex Blanket.

Maria Selby, the company's CEO, told me it was because the company wanted to 'initiate a new kind of conversation around period sex, one that encourages people not to feel shame or shy away from the human body during menstruation. This is more than a functional blanket, this is another opportunity to bust through yet another period taboo and to initiate a much-needed dialogue about period sex, and sex generally.'

She added:

> The aversion to period sex likely comes from generations of systemic sexism and stigma that teaches both men and women to feel uncomfortable talking about menstruation. It's up to us to change the trajectory of thinking that things like periods and period sex are gross, when they're really just natural parts of human existence.
>
> So many people are made to feel afraid or ashamed of having sex on their period, but the truth is that period sex is totally safe and natural. In fact, orgasms release hormones that can help ease the pain of menstrual cramps. We wanted to start a conversation about period sex, and figured that this product was a great way to do that.

Coming in just shy of £300, it's a fairly expensive investment when an old towel will do the same job. Admittedly a crusty beach towel probably isn't as soft as

Thinx's purple large satiny number. Nor does it boast the company's signature 'four-layer absorption technology' first shown off in its knicker line, designed to absorb the mess of bloody sex in a discreet and efficient way – protecting your bedsheets. But nor does it require immediate tending to in your languid post-coital state; the cleaning instructions on Thinx's period sex blanket are somewhat of a mood killer. To take care of your Thinx Period Sex Blanket, you are instructed to rinse it immediately on a cold wash on delicate cycle, then hang up to dry away from direct sunlight. Who can be bothered to rinse anything immediately after use in that kind of activity?

While Thinx's intentions are laudable, perhaps there is something slightly perverse about buying an expensive product to do something that, if you want to do it, should be perfectly normal, natural and free.

Maybe a special blanket makes period sex seem like a weird fetish you keep a magical cloak for under the bed. As I previously noted, it is a kink for some, but for the majority who do already indulge in red sex, all that's required is passion, confidence and, if you remember in the heat of the moment, a sizeable towel you care not a jot for.

I hope by doing a deep dive on period sex together I have blown away some of your own preconceptions. Or, if you really are very comfortable with sex during your flow, made you stop and appraise why you feel the way do. Or if you loathe it, distinguish why. Is it because you feel vile during your period or you are squeamish about blood generally?

Chances are, most women have inculcated so much shame about their bodies during menstruation, that flipping it on its head to the point where you could give yourself permission to feel sexy instead is a tall order. Moreover, men, who are taught so little reality about periods in the first place, also have to get over their fears and prejudices if they want to satisfy their needs and their lovers' needs during their periods.

But it can be done. With the right partner, unless they have a genuine case of hemophobia (a fear of blood), you can screw the shame away. If you feel like doing the dirty, do it proudly and in the knowledge that it's safe, good for you and nothing to be ashamed of. Period sex isn't some filthy weird kink reserved for the depraved in society who have a particular penchant for rare steak and dungeons. It's just sex with a bit more mess. As women, we may be at our horniest at that time of the month and may have been denying ourselves the release of menstrual sex because, to quote my friend, we are so worried about our partners 'finding it gross'.

Period sex could be the best action of your life, and old school views about what's proper may have been holding both you and your lover back. If you are even just a little bit curious about it, I urge you to try having sex on your period. And talk to your partner first. Clear the air and reassure them about any of their fears. Some guys genuinely think we bleed buckets and it's like a murder scene down there. Admittedly, some days are heavier than others – but equally, it's not like a gushing river in our pants. Or a red tap turned permanently on high.

As women, it's only right that we do lead the way on period sex. It's our bodies having a moment and we need to lead by example. That means we have to feel confident in our horniness and desire for sex at that time so that our partners feel aroused by our arousal, emboldened and like they have our full consent to get down to it.

I fully recognise that some of the women reading this would never want anything other than a tampon going near their vaginas during their period. They can't even bear to be touched during their flow. All they want is a hot water bottle, space and their favourite snacks. Lots of snacks. I fully salute this.

But it's still important for all women that period sex is normalised and recognised as a perfectly healthy and clean activity that women want to engage in. It's not the preserve of sexually ambitious men who wish to demonstrate their masculinity. Nor should a period mean a woman feels compelled to literally bend over and offer an alternative hole.

Because if we do manage to normalise period sex, there is a wonderful, almost magical, much bigger win for women the world over: there will not be a day in the month where women feel like they ought to loathe their bodies. Like they are less sexy and dirty in some way due to a perfectly natural bodily function.

Of course, all body dissatisfaction can't be bonked away through period sex. I may be a believer in change but I ain't that naive. However, removing a monthly cycle of shame would be a start. And if more women felt like it wasn't icky or weird to want to act on their horniest desires with another should they strike during their period, how cool would that be? How transformative could that new feeling of permission be to their own self confidence, self-worth and their relationship with their lover?

Make no mistake, any partner worthy of loving will find your openness and willingness the biggest aphrodisiac. Of course, they have to be up for it too. Consent must cut both ways and if, for some reason, they can't get on board the period sex train, do not take it personally. It isn't your fault. Or your body's fault. It's their own deeply ingrained issues. Don't forget that they have been privy to the same millennia-old influences you have, telling the world that women are dirty on their periods and should be excluded from society. If they can't shake off these nonsense, religiously infused notions, please don't inhale it as your problem.

Instead, run a bath, pour yourself a glass of your favourite tipple, and sate your own desires – feeling stunning, soapy and liberated.

Shame is a deeply ingrained worm. It burrows deep and can take years to locate and banish. It cuts across both genders in different ways. But if you manage to happily have period sex for the first time because this chapter has made you consider what's actually been holding you back and reassured you of your concerns, I am one satisfied lady.

CHAPTER ELEVEN

WANTED AND UNWANTED BLOOD

'I still can't help but pray that this month is the last.'

Anonymous 18-year-old trans guy from New Zealand, writing on Buzzfeed

Up until this point we've only spoken about periods affecting women. And it being the norm. But how do periods affect those who can't have them but want them? Or those who have them but don't want them?

I had my own experience of deeply unwanted blood of course, during the two years we attempted and failed to fall pregnant. It's a lonely and terribly dark place trying for a baby. Your period takes on a whole new persona. It becomes the enemy, the terribly sad and unforgiving bearer of crushing disappointment each month. You learn to hate it. Even just the hint of it in your pants.

So vivid is the moment it comes and takes away all of your tiny shreds of hope that this month *will* be different, I can remember exactly where I was when each of those twenty-four periods began. In the toilet down the corridor from my studio multiple times just before going on air; in the airport before boarding a flight home post a New Year's Eve trip; in a friend's loo during her one-year-old's agonising birthday party. These are just a few of the memories of my unwanted, hateful periods burned onto my brain, and, most of the time, I would have to emerge from said toilet as if everything was fine, blink back hot tears and smile outwardly, while inwardly damning my stupid, broken, barren body.

The idea of someone wanting, even desiring a period during my normal mental state is an anathema to me. The idea of it during those two years of mental and physical hell? It wouldn't have computed.

It's time for you to meet Jen Palmer, someone whose story I have never forgotten.

WANTED BLOOD

Even fourteen years later, I can still smell the smoke of the raging campfire as I sat with a relative stranger under the stars of Swaziland's dramatic skies, as my brain raced to process the startling piece of information that I'd just been given. The girl next to me was Jennifer Palmer, an utterly beautiful strawberry-blonde and the best masseuse in my new travel group (she'd just given me a very generous shoulder rub), and she'd just explained why she was trekking around South Africa on her own.

'I don't have a womb,' she'd told me, after a deep intake of breath. 'I was born without one and I only recently found out.'

The information dropped like a stone between us and I was momentarily dumbfounded. I didn't know such a thing was possible, and from the haunted and almost embarrassed look on her face, neither did she. Until she did.

I quickly regrouped and found the words to gently ask the next obvious questions: 'What did that mean?' 'How did you find out?' 'When did you learn this?' 'Do you have periods?' And the gulp-inducing one: 'Can you ever have children?'

She stumbled through her answers, which she was only just learning the script for herself, as the diagnosis was terribly fresh.

In a moment, I will share Jen's full story, because it is one of bravery, but also because it tells of a rare thing: wanted blood. A longed-for period. A hankered-after reproductive system with all the ups and downs that brings.

However, I would be lying if I were to say that the details of her story are drawn from my memory of that conversation all those years ago, when I travelled on a pleasure quest across Swaziland, Botswana, Mozambique, South Africa and Zambia as a student with a crew of people who became firm friends.

The truth is, Jen, did tell me a lot that evening around the fire, her eyes gleaming with smoke and emotion. She chose to open up to me about her condition during the most painful chapter of her life. And I didn't know it at the time, but I was one of the first people outside of her immediate family that she told of her diagnosis, zealously guarding the information about her insides that made her question her outsides.

Despite her candour, all that I remember of that seminal exchange is 'I don't have a womb.' I can still hear her saying it to me. And vividly recall my own breath being taken away.

I have thought about that conversation so many times over the years. Jen and I lost touch but remained friends on social media, meaning occasionally I'd see a stunning shot of my Africa travel buddy cuddling a mate or posing in front of suitably pretty holidayscape and I knew she at least looked like she was doing well after her brutal awakening. It didn't stop me wondering, though, how she was really coping away from the lens.

To be born as a woman without a womb – something that we are told is so fundamentally female – it goes without saying that it's a major body blow to process. Especially if you only found out at eighteen, the age you supposedly turn into an adult and are ready to face the world with the cards you've been dealt. Even if you had doubts about wanting children, a woman presumes she has the choice – should she change her mind.

Most people I've spoken to about periods during the course of writing this book don't want them. If they could flick a switch and turn off the pain, the hormones, the mood changes, the blood, they would. Some of them have. But Jen is in the rare camp of desiring, even coveting, a period. I knew I had to get back in touch with her and pick up where we left off fourteen years earlier.

Neither of us really had the words back then, but she's now *more than* found her voice. In fact, it's a roar, laced with occasional vulnerable dips and audible sadness. This, in her own words, is what she told me:

> When I was seventeen, I hadn't started my period. I didn't think it was that weird, but I also didn't bring it up with my mum, as I was uncomfortable talking to her about that stuff. I pushed it to the back of my mind and tried to carry on without thinking about it.
>
> I woke up one morning with some pain, and while it turned out to be unrelated to what would become my diagnosis, it did cause me to go to A & E and led to me having some ultrasounds.
>
> This, in turn, led the doctors to seeing some stuff they weren't expecting, but ultrasounds aren't always that clear. All I was told at that point was I needed an MRI scan. I had one and went off to Nottingham University without the results. Returning home a few weeks later, as I didn't like my course, had left uni and was intending

to reapply somewhere else, I went into the hospital with my mum for my results.

Now aged eighteen, I was sat in a big waiting room and only have a few memories from this day. I remember seeing a list on the wall of patients seeing the junior consultant and the senior consultant – and then hearing the reception saying that the head consultant wanted to see me.

That's when I had a real moment of foreboding.

I went in and she opened the file. Barely looking up, she said: 'There does seem as though there are some issues. Your uterus is missing. Your fallopian tubes are partially formed. And the top two thirds of your vagina is missing.'

I was eighteen.

I looked at her and said immediately, 'So I can't have children?' That's all I remember caring about – not what was missing, but what I couldn't do. I didn't care about any missing parts at that point.

She shrugged her shoulders and said, 'I don't know what you want me to say.' The doctor carried on reading out the report and using medical language. It was clear she didn't know what to do or say. She also told me my condition was extremely rare and affected one in 50,000 women. I later learned this was far from the truth – it actually affects one in 5,000 women.

My life was falling down around me in that hospital room. I remember she then performed a vaginal examination which felt very invasive and I am not sure if it was even necessary at that point. Finally, she gave me the name of the condition I had only just discovered I had: MRKH (full name Mayer-Rokitansky-Kuster-Hauser syndrome, named after the doctors who discovered it). A broad-brush definition means the reproductive system starts to grow but doesn't fully develop.

I don't remember much about the appointment after that point. I have had a lot of counselling since then and had to relive it. The doctor could have handled it far better and it was a real moment of trauma. The only other clear memory I have was of leaving the appointment room, with my mum following me out as I kept repeating out-loud, 'I can't have kids', and my mum turning to me, crying, saying: 'I just wanted you to be perfect.'

She meant for me – but the implication was that you aren't. That you aren't a proper woman. She didn't mean to hurt me, but her words did and they've stayed with me. My mum really didn't know what to say or do. No one knows how to talk about this stuff. You just don't feel like you work.

I also remember feeling very confused as to how this could happen. It sounded very science fiction – not to be born with something you thought you had. A 'rudimentary uterus', what did that even mean? It turned out that my womb had started to form but hadn't finished. It made me feel disgusting and malformed – like something should be removed from me.

Jen has had a long time to reflect on her condition since that appointment. She's met many other women living with MRKH, joined support groups, run ultra-marathons to raise money and awareness for the syndrome and has even appeared on a BBC documentary about fertility – travelling to Sweden where the first womb transplant successfully took place.

She's also been relatively lucky that her vagina responded well to the necessary dilation treatment – which is about as much fun as it sounds (in her case, go to the hospital twice a day for two weeks, while different sized medical dildos are inserted to stretch your vagina for the purposes of sexual intercourse). The nurses called her their 'star pupil' in a much needed-moment of black humour; she's never needed to return – whereas some women have to repeatedly go in for stretching each time they want to have sex.

Understandably, her inability to naturally bear a child, in spite of having ovaries filled with eggs, has preoccupied her mind. Especially after she turned thirty and her friends started having children. As of yet, Jen hasn't actually met the guy she wants to create another human with, but she has made peace with trying to have kids either via IVF surrogacy or adopting. She describes not being able to have children as something which has made her personally 'feel less like a woman'.

'A big part of our identity as women is being able to have children – if you can't or don't at least have the option, then what are you good for? That's a question I live with. Having MKRH has always made me feel I have to be very successful at whatever I do – as I am always making up for something,' she explains, with a refreshing candour.

And she's been highly successful at whatever she's turned her hand to, whether it's smashing those ultra-marathons in her spare time or in her day job running the business development at a charity which helps women and children.

But interestingly, it's only recently that the reality of never having a period has fully dawned on her, as Jen has fought to protect herself from desiring things she simply cannot have.

Not having a period is usually the 'first tell' of her condition with new boyfriends. The fact she never 'comes on' becomes a question for them; something they notice as being different about her. 'One of my first boyfriends got suspicious that I never seemed to have a period and he looked up why certain women don't bleed online and came up with MRKH. And then confronted me about it. It was tough as I wasn't ready to tell him yet,' she confides.

Periods aren't just an annoying giveaway for Jen though. For her, periods represent part of being a woman she will never experience; a teenage rite she never went through and another club she can't join.

While Jen opened my eyes and mind when she told me she didn't have a womb all those years ago, she's managed to do so again in our latest conversation with another statement I've never heard anyone else utter before:

'It would be so nice to have a period.'

You, or those close to you, may not suffer with their periods. You may choose to avoid having one altogether through a particular kind of birth control. You might never have even thought much about your monthly flow, other than when you are buying some tampons (and reading this book), but I bet you an entire year's supply of pads that you ain't ever felt lucky for having one. Unless you really thought you might be up the duff that month and didn't want to be – then you'll practically kiss the dirty sanitary bin in the public loo where you find the amazing info that you're not.

So, consider that, to some women, a period is a beautiful sign of health and something they dream of being able to have. It is very much wanted blood, which, as Jen puts it, women in her position have been 'denied'.

'It would be so novel to bleed. As I've dealt with the pain of not being able to naturally have children, the feeling of being left out of a club has dawned on me; it's a club most women are just in. It's about being able to participate,' Jen admits ruefully.

'In conversations with other women, they moan about their periods; they ask me for a tampon; they complain of having to do the pre-holiday Boots run to buy sanitary gear or not being able to have sexy time because they are on. I cannot have any of these conversations or give anyone a tampon. I just have to say 'that sucks' and move the chat along. I feel like an outsider.'

Jen does still have a few period symptoms because she has ovaries and the accompanying joyous hormones. She talks of being hormonal once a month and 'feeling my body have a cycle'. Very occasionally she gets bad cramps, which she believes is when she can feel her ovary releasing an egg. The cramps are abdominal and she describes feeling blocked. Because her womb is part formed, her body wants to shed something but it just can't flow.

Being an all-round top woman, Jen tries to have a good sense of humour about her lack of monthly blood. 'I save a lot of money on period products and can go swimming or have sex anytime I like. Plus, I'm sure after one period, I'd hate it like most other women.'

She's also thankful that her 'scars' from MRKH are not visible like some people's: 'I am lucky that I don't wear my scars on the outside. People don't look at me in the street and know I am missing a womb. I haven't lost a limb or anything people can see, but as I get older and consider my mortality, it's dawning on me more and more that I will never have a period and it's so weird.'

Not having a period has made Jen feel broken and left out of the women's club – but what if *having* a period had the exact same effect? What if having a period made you feel out of sync with your gender and whole identity?

UNWANTED BLOOD

Meet Cass Bliss, an American twenty-six-year-old non-binary trans activist, and the creator of the website and hashtag #bleedingwhiletrans and the Instagram account 'The Period Prince'.

Cass does not identify as a man or a woman and only came out as trans a couple of years ago, after a very traditional upbringing in the bosom of a loving, highly conservative Baptist Missionary family. Because Cass is non-binary (which describes anyone who feels they do not exclusively fit the accepted definitions of man and woman), I will refer to Cass with the pronoun 'they' as opposed to 'he' or 'she'.

It was in July 2017 that Cass came to people's attention around the world after posting a photo of themselves free bleeding into their trousers with a sign saying: 'Periods are not just for women #bleedingwhiletrans.'

The image caused quite the storm. And while Cass, a self-professed period activist, is fighting for the world to recognise that it is not only women who have periods, for the tampon companies to stop branding everything girly pink and alienating menstruators like themselves, and for male toilets to have sanitary towel bins, I was more intrigued to learn about living with unwanted blood, month in and month out. What does that really feel like and what does it do to your identity?

While the image of Cass bleeding all over beige chinos on a park bench is certainly arresting, it was the poem posted alongside the post that caught my eye more and captured my imagination:

Y'all know I'm trans and queer,
And what that means for me all around,
Is something that's neither there nor here,
It's a happy, scary middle ground.

So when I talk gender inclusion,
And I wrote these rhymes to help you see,
I'm not tryna bring up something shallow,
Periods are honestly pretty traumatic for me.

See my life is very clearly marked,
Like a red border cut up a nation,
A time before and a time beyond,
The mark of my first menstruation.

So let me take you back,
To the details that I can still recall,
Of the day I gained my first period,
And the day that I lost it all.

I was 15 and still happy,
Running around, all chest bared and buck,
Climbing trees, digging holes,
And no one gave a single fuck.

I mean I think my ma was worried,
So I went and grew out my locks,
A sign I was normal, still a girl,
A painted neon sign for my gender box.

So, the day I got my period,
My god, a day so proud,
This little andro fucked up kid,
Had been bestowed the straight, cis shroud.

The relief got all meshed up in my pain,
In that moment, I sat down and cried,
Just thanking god I was normal,
While mourning the freedom that had died.

Everyone told me my hips would grow,
I looked at them and couldn't stop crying,
'What's wrong with you? You'll be a woman!'
They kept celebrating a child dying.

See my body had betrayed me,
That red dot, the wax seal,
On a contract left there broken,
A gender identity that wasn't real.

Most people deal with blood and tissue,
And yet my body forces me to surrender,
Cause every time I get my cycle,
Is another day I shed my gender.

My boobs betray me first,
I feel them stretching out my binder,
I send up questions, 'am I cursed?'
And wish to god that she was kinder.

The five days it flows,
I try to breathe, I dissociate,
While my body rips outs parts of me,
Leaving nothing but a shell of hate.

The blood drips from an open wound,
Of a war waging deep inside my corpse,
The battle between mind and body,
Immovable object; unstoppable force.

There's some pretty brutal sentiments in this unique ode. Cass describes periods as 'traumatic'; a way their body 'betrays' them and as a 'war waging deep inside'. Their first period was the day they 'lost it all'.

You might find this terribly hard to process, but if you think about it, talking about your period is still a major taboo when you are a woman who accepts a monthly flow as part of your natural fate. Now imagine not wanting a period because you don't identify as a woman, not being able to stop it and wanting to shout from the rooftops about it.

Periods for this group of menstruators are uber-taboo. In fact, Cass is one of the only people I could find who had put their name to this problem. The internet is scattered (albeit sparsely) with anonymous posts on the subject, but mainly from folk going through a whole world of pain, like this eighteen-year-old trans guy from New Zealand, writing on Buzzfeed: 'As a trans guy it's hard. Scientifically I know until testosterone injections kick in, it's gonna happen, but I still can't help but pray that this month is the last.'

Cass is still battling with periods when we catch up over the phone – this is despite a course of testosterone injections, which have now stopped due to health insurance complications. If anything, testosterone injections made Cass's periods even more aggressive rather than ending them.

During our conversation, Cass takes me back to that first period. The family base was in the Congo as they worked as Baptist Missionaries. Gender identity, periods and transgender rights were not topics of conversation in a highly conservative environment. Cass recalls:

> I was raised with missionary kids. We didn't have internet access or access to pop culture. We didn't have smartphones or any ability to read up about gender issues. I lacked the language for how I felt or to express myself as trans.

All I knew was I didn't want my period because I didn't want to be a woman.

> Up until my first bleed I felt androgynous. I had two older brothers and didn't feel a huge difference to them. But once my period started when I was fifteen, everyone started saying I was now a woman – which I didn't want to be. Suddenly I felt I had to follow the rules, even though everyone knew I was different.

Cass went on to study queer theory at college, which finally gave them the language and the insight to express how they felt within themself. It also opened Cass's eyes to a community which had existed all along their childhood but now they could access. Cass duly came out as non-binary trans and the period activism, under their fun internet moniker, 'period prince', soon followed.

But finding the language to express how they felt still didn't change their monthly biological fate. Cass explains:

> Gender dysphoria is when your body doesn't match up to how you feel it is. For me, my breasts are the biggest area where this is true and my period makes it worse. Normally I bind my boobs so that they aren't as visible under clothes. But when I get my period my boobs swell. So, then I can't wear the bind. So, I move to a sports bra and then I get misgendered a lot more as a woman, because my boobs are suddenly visible and people then start seeing me as female.
>
> This then throws up all sorts of issues when I have my period like, am I now a boy with boobs? Should I use the men's room or not? The men's room doesn't have sanitary towel bins but I get funny looks whichever I choose.

Cass's period is an unwelcome reminder of parts of their self they don't want to remember or be defined by. Their womb and their breasts.

What Cass battles with every single month is the widespread mindset that a period makes you a woman, which is something Jen has also grappled with. Not all women menstruate, and it doesn't make anyone less of a woman for not doing so. It's not like the minute the menopause hits, women suddenly stop being full women. Equally, as in Cass's case, not all those who menstruate identify as women.

When I ask the self-styled period prince if they would get rid of their period if they could, there's a pause and then a definitive response:

> Ideally I would love to get rid of it. Personally, I get a lot of dysphoria because of my period. It makes me feel uncomfortable in myself. It represents a battle within me. It's a marker of what I am not. My period makes me feel wrong. It makes me consider the question again and again: 'Am I a freak?' But I am not sure about a hysterectomy as it's very invasive. A lot of trans men do live with having a period that they don't want. It isn't easy but it's their reality.

Cass's blood is very much unwanted and makes them question who they are. Their period is painful in a whole other way. It makes them doubt their very identity as their unwanted breasts swell and then society makes them feel like the one thing they're not: a woman.

Jen's lack of period blood makes her question who she is too. It has made her feel left out of 'the woman club', and like she needs to overachieve as if to somehow make up for the fact she was born without a womb and therefore the ability to menstruate and procreate. Her absent blood is very much wanted.

Periods run deep, and they can be powerful signifiers

of health, fertility and identity. But we have been so busy making sure no one talks about them at all, we have failed to see the other types of pain, beyond the physical, they can cause on a daily basis.

To quote the period prince in one of their latest YouTube videos: 'Not all women menstruate and not all those who menstruate are women.'

CHAPTER TWELVE

NO BLOOD

'I like women on the other side of periods. They are tough. They are reinventing themselves and shrugging off all their previous shit.'

Miranda Sawyer, journalist and broadcaster

You can spend years wanting it. The first show of browny-red in your knickers just to prove you are 'normal'. And a grown up. If you're struggling to recall this moment, or this wasn't your experience, then you need only delve back into the seminal *Are You There, God? It's me, Margaret* by Judy Blume, which effortlessly transports you into the mind of an anxious teenage girl who craves her period and bigger breasts as confirmation of her normality.

And yet, once your period starts, it's pretty grim, annoying and – even if you don't experience much discomfort – a faff.

But then it stops.

Another rite of passage begins: the menopause. And guess what? Some women *miss* their periods. They mourn them. For all sorts of reasons. That's what I want to explore now, as we near the end of our red quest together.

What is there to miss about one's monthly bleed? The whole concept of any benefits to a period (apart from the obvious medical ones linked to fertility) is a total anathema to me. I suspect it's the same for anyone else with a period condition like endometriosis or anyone who experience heaviness and painful cramps, month in and month out.

But there are plenty of women out there who *do* miss their period – or at least elements of what it brought to their lives. It would be misguided of me to think that everyone at the end of their period careers throws a massive party, using up their final batch of tampons as swollen stirrers in celebratory Bloody Marys.

I already identify with those women who see the end of menstruation in their lives as one of the most liberating moments. My mother and mother-in-law both fall into this camp, both having suffered debilitating period pain all their bleeding lives.

Personally I am convinced that my mother and grandmother had undiagnosed endometriosis. It is genetic and their symptoms were extremely similar to mine. Therefore, it's hardly surprising that the three words my mum used when I asked her to describe how she felt as her periods stopped were: 'joyous, free and liberated'. This is the woman who begged her gynaecologist for a hysterectomy and wasn't granted one on the grounds of it being a major procedure she didn't desperately need.

She confided, in her sweetly handwritten responses to my typed questions (computers and my mother have yet to collide): 'Having periods never secured my femininity. But now I am a pain-free woman.'

Despite my mother and mother-in-law loathing the hot flushes that came with the menopause, which are still plaguing both of them long after their periods ended, they were both so grateful when the bleeding was finally over. 'Relief' was the word my straight-talking Swedish mother-in-law used.

Dr Helen Pankhurst, the feminist activist and great-granddaughter of suffragette leader Emmeline, explained that she felt similarly. Her periods ended rather abruptly following a hysterectomy and treatment for endometrial cancer. Before getting ill, she described her relationship with her periods as 'problematic in the extreme! Often painful and a bore, at other times incapacitating; I would faint and just have an awful time of it – this was how it was before it got much worse due to the cancer.'

I asked Helen for her take on periods because she's done a great deal of thinking, campaigning and writing about the modern-day woman and her lot. It's no surprise, considering her eminent family heritage in women's rights. She's also become a great friend to me over the years, after we met through her campaigning. In fact, I think of Helen as one of my ultimate women (I mean, she's a Pankhurst, after all), so I rather naively presumed that she would have something positive to say about her period now she's safely on the other side.

But no. The opposite. 'I just felt worse [when on my period]. Periods have fuelled my feminist anger. Why, oh why, do we have to suffer in this way? And why does society add layer after layer of taboos to make it so much harder?' She also said she no longer felt 'incapacitated' once she stopped menstruating and was totally 'liberated'.

Florence King, the American Conservative novelist and essayist, went further and put it like this:

> A woman must wait for her ovaries to die before she can get her rightful personality back. Post-menstrual is the same as pre-menstrual; I am once again what I was before the age of twelve: a female human being who knows that a month has thirty days, not twenty-five, and who can spend every one of them free of the shackles of that defect of body and mind known as femininity.

While I wouldn't agree with this entire sentiment, I do feel shackled and as if my period is a defect. I do feel I am still able to be free, but I am also intrigued by the idea of who your period makes you and whether the monthly cycle brings out different versions of a woman's self.

Florence King obviously relished the ending of monthly seasons; the up and down of a hormonal rhythm. Similarly, I won't miss the knuckle-dragging gremlin version of myself who rears her head when I menstruate naturally (without masking it with the aid of the pill or heavy painkillers). The version who has no patience because of the pain, can barely put one foot in front of another, and growls at anyone nearby with energy and vim. And for whom a giant hot water bottle becomes as necessary as chips and cheese at the end of a long drunken night out. (Must absorb. Must absorb.)

But for others, it's an entirely different story. The ending of a period induces a sense of grief for what they will lose and trepidation about what is to come.

Miranda Sawyer, the journalist and broadcaster, found herself compelled to pen a book about ageing as she headed towards her fifties and endured a mid-life crisis. The resulting tome, *Out of Time*, forced her to assess who she is and where her life is going.

At the time of my writing, Miranda is in the process of stopping her periods and heading into menopause. This is how she explains how she felt as she reached the end of her period career, and experienced four months with no blood:

> I just felt flat. The same all of the time. I found it quite boring. I won't miss my periods in the sense of the actual blood. Nor do I worry about them ending and what that means in terms of my life. But what I do worry about is being flat. About my hormones staying the same the whole time and no longer being affected by a cycle.
>
> I enjoy the difference of being weepy one day and angry another. My cycles present versions of me. There are days you feel hornier than others. I like the anger, sadness and rush of blood to the head. My anger fuels

> me. Or my sadness. I enjoy all of those feelings. I don't want to feel the same all of the time.

Most women do experience a whole range of feelings in the run-up to their period and this emotional range is something one can miss on the other side of menstruation.

Of course, you might be the sort of person who would relish never having unpredictable peaks and troughs – but Miranda makes a compelling case. She's keeping a close eye on her mood as many women fall into a depression when they hit the menopause and are offered Prozac as well as HRT. (She says she will take the drugs if she needs to.)

Interestingly, it can be argued that the ebbs and flow of a monthly cycle may actually give a woman the psychological permission to break out of the normal 'feminine' polite behaviour expected of her by society. Amanda Laird, the author of *Heavy Flow: Breaking the Curse of Menstruation*, buys into this idea and relishes this 'opportunity' for women: 'We can use PMS (premenstrual syndrome) as an excuse to break out of what's expected of us – being all girly and nice. During the month we can suppress our emotions and then in the run up to our period, we can't any longer. Our emotions bubble up and cannot be contained. This can be liberating.'

Our periods as mood liberators is a completely different way of thinking about them that I'd never considered before. Or, as the writer Christa D'Souza puts it: 'When your period arrives you have that moment where you realise "Ah. That's why I was being a c**t".' Punchy.

Christa, a mother of two, penned a book on the menopause, *The Hot Topic: A Life Changing Look at the Change of Life*, detailing all the shifts that take place as one stage ends and another begins. When I ask Christa about how she feels about her periods ending, she swiftly replies with one word: 'nostalgic'.

'I miss the ritual of it. The relief I'm not pregnant,' she quips.

But it is the rhythm and the release of it she misses the most: 'Every month it felt like ablution; getting rid of something down there. I loved that feeling. Getting all the stuff that shouldn't be in your body out.'

Like Miranda, she also enjoyed the pressures and the build up to it and uses a powerful comparison that will stay with me a long time: 'I liked the momentum of having a cycle. Ten days or so after your period you would be at your peak with the most amount of oestrogen in your system. But now, since the menopause, it's like never having a weekend – where all of the weeks blur into each other.'

Never having a weekend again, once your period ends. *Wow.*

The rhythm, the cycle, the mood changes, a more liberated version of yourself – these are just some of things women miss or would miss

about having periods and I find it totally eye-opening.

Amanda says she would also miss the sense of release but also renewal. Saying a sentence I could never manage uttering, unless dwelling in a land of dark opposites, she chirps happily: 'I love having my period. That release feels great. I always feel fresh afterwards like I am starting a new month and I can put that last cycle behind me.' Though she is quick to say that she knows that it's a sentiment that would sound like 'bullshit' to anyone who suffers with their periods each month.

The sexiness and youth associated with periods are also states Christa mourns:

> For me, I love that feeling of sexy just before you got it. You felt alive and sexy in a way you don't quite again. Having my period also made me feel younger. It marks a passing of time. Just before it ends and you are perimenopausal, your final periods get heavier and closer together. It's like you are a Roman candle burning down. I think this is your body attempting to sprog a final time. But ending your periods is sad and, just like starting them, it's a rite of passage women go through. Men don't have this. They just get saggy bottoms.

With that farewell to youth can also come a farewell to a form of 'period sisterhood' you may not have even noticed you are a member of. I am immediately minded to think back to Jen, whom having been born without a womb, felt so very excluded from this. And for all my period dramas – I had never even appreciated I was part of something bigger. Christa explains:

> There is no sisterhood with menopause unlike periods. Menopause is officially one day, twelve months after your last period, and then that's it. With periods, all women have good leaking stories. I remember a woman I didn't know tapping me on the shoulder when I had leaked through some white trousers telling me about it. Periods can be a bonding thing between women in a way menopause just isn't.

Periods for other women are a vital sign of health and their passing means saying goodbye to one of the body's signals that all is well – or isn't. That's certainly how Louise Chunn, founder of therapy platform welldoing.org and former editor of *Psychologies* and *Good Housekeeping* magazines, feels:

> We don't know very much about our insides and how they work as women. Until I had my first baby, it's all a slight mystery what's going on in there. You don't know if everything is working as it should and then your period arrives and I recall it giving me pleasure and a sense of relief. Because you think 'Ah, OK. Everything is working in there'. Periods made me feel fine. I didn't think of them as a curse.

Earlier in the book, we met Dr Jane Dickson, an obstetrician and vice president (Strategy) of the Faculty of

Sexual and Reproductive Healthcare, who believes women don't need their periods other than for fertility reasons. When I challenged her by asking couldn't periods and how they manifest be a key indicator if something was wrong health wise with a woman, she stuck to her guns and said there would always be other signs – so, no.

Lara Briden, a New Zealand-based naturopath (a holistic practitioner who uses herbal and natural remedies) couldn't disagree more. The author of *The Period Repair Manual*, Lara believes your menstrual cycle is a 'monthly report card for good health'. For her, a monthly bleed is not the important part of having a period. The period is a side-show; the tax women pay for having a natural cycle. For Lara, it's all about having a natural ovulation cycle so women can make hormones – namely natural progesterone and oestrogen – which she calls 'deposits into the bank of long-term health', bringing benefits for bone density and one's immune system. She says, 'For most women with no menstrual conditions or illnesses, periods should be a small event we can manage.'

In terms of contraception, this is why she favours the hormonal IUD, as women still make their own hormones, as opposed to the pill, where women bleed but do not have a natural cycle. She believes the pill and the like are akin to the chemical castration of a woman and that future generations will look back on doctors' current advice to teenage girls struggling with periods to take the pill with incredulity – that they'll question why on earth we were telling young women to stop themselves from experiencing their natural hormones whilst they were still growing and changing. This school of thought has gained more traction of late away from the medical community and is partially behind the rise in more young women using apps such as Natural Cycles as a form of contraception, or period-tracking apps like Clue as a way of understanding why they feel certain ways at certain times of month.

Lara makes a compelling case and I am not qualified to prove or debunk her theories either way – but as someone who has benefited from being on the pill (in terms of helping me cope with periods as an endo-sufferer and also as someone who has successfully had IVF and benefited from hormones artificially flooding my body), I love modern medicine.

Acupuncture and herbs did nothing for me when trying to get pregnant. Nor have they helped manage my monthly pain. Only cold hard drugs make any difference. And playing with my hormones so I produced a load of eggs like an overactive hen for my husband's sperm to remotely fertilise in a little dish happily produced our son.

But that's not to say women who can naturally have their periods and monthly cycles shouldn't treasure and appreciate them. And that's what Lara will miss about her period: the natural deposits of progesterone and oestrogen her natural ovulation cycle delivers each month.

What I can confidently support Lara on is the fact that

modern medicine, despite all of the incredible leaps and bounds it has made in the realm of fertility, still knows shockingly little about women's menstrual systems – particularly about what causes the problems and how to cure them. For instance, nobody knows what causes endometriosis and there is still no treatment to rid the body of it. That shouldn't be the case for a condition that affects approximately 5–10 per cent of women and girls (176 million) around the world – and that's if they get diagnosed at all.

But whether your periods have been grim and gruelling monthly occurrences, or they've been just fine, you could still miss them when they are gone. And in a chapter dedicated to what certain women miss about their periods, I couldn't omit the biggie: fertility. The answer a woman gives when asked how she feels about her periods ending may depend on whether she has had kids.

Let me issue the obvious caveat at this point – many women who do not have children never wanted them and are absolutely A-OK about this decision. However, for those women who stop bleeding without having had children, who wanted them badly and it never happened – either because of medical issues or they never found the right person – it's a major moment. Or for mothers without children, those who lost babies at birth or later, the stopping of a period marks the end of the possibility of new life.

The window of chance and possibility to reproduce again has closed – that chapter of the woman's life is over.

Men don't have an equivalent biological moment.

Sadness, bitterness, anger, grief are just some of the emotions women in this large boat report feeling at this point.

The writer and actress Lena Dunham sparked debate when she wrote a searingly honest piece for American *Vogue* in 2017 about her decision, aged 31, to have a hysterectomy. She chose this path to end years of period-induced anguish as a long-time sufferer of the harshest form of endometriosis.

I am not interested in debating whether she should have shared it or whether it was responsible to do so. People who genuinely believe that others sharing their experience of difficult, brutal things will lead to an epidemic of copycat cases, are as foolish as those who think photos of obese models will lead an epidemic of young folk into eating themselves into a stupor. Humans may be impressionable but we also have free will and brains. Of course, some people may be influenced, but the worry over a sudden hysterectomy epidemic (which I've seen expressed on TV and in print) is a major and highly illogical overreaction.

Lena's piece stopped a pregnant me in my tracks as I brewed a tea in the kitchen at work during one of my final shifts before going on maternity leave. Not only was it stunningly written, but as a fellow endo-sufferer who had managed to get pregnant, the final two paragraphs moved me to tears and, I say this in the least smug way possible, hug my protruding stomach with sheer gratitude and guilt:

> The children who could have been mine do break my heart, and I walk with them, with the lost possibility, a sombre and wobbly walk as I regain my centre [...] Adoption is a thrilling truth I'll pursue with all my might. But I wanted that stomach. I wanted to know what nine months of complete togetherness could feel like.

That's what the end of periods meant to one woman who had just had her very ill womb removed. The end of the chance to have 'nine months of complete togetherness'.

The end of periods for some women can be bloody devastating and understandably so. Knowing one's body can produce, house and birth another, and having never been able to take it up on that capability, while most women around you have, is one of the most painful realities. By the time a woman in this position finishes her periods, she may be in a much better, peaceful place, but the official ending of her chance can retraumatise and finalise something that hadn't been final until that point.

While I will be thrilled to be rid of my periods (and I still don't know whether I will have to take action to end them myself earlier than nature's schedule like Lena), I will forever be grateful to them and respectful of them for allowing me to grow and birth our boy.

The end of fertility, the closing of that window, is a poignant time for many women. The stopping of periods was a sad moment too for Louise Chunn, even though she's had children and a miscarriage, because, as she put it: 'I left the company of fertile women. And went to the other side.'

She remembers going to see a gynaecologist who confirmed her change of status and delivered a sharp reality check – in that lovely unemotional way doctors often do. 'He did an ultrasound and confirmed I was menopausal. He said my left ovary was atrophied [wasted away]. And my right one too. I remember that moment very clearly. I felt sad and like a deflated balloon. And thought, "Right. That's that then". I didn't want any more children, but it was still a moment all the same.'

And, on that point, I think all women – regardless of how they have felt towards their period – can relate to the idea of the end of their menstrual cycle being a moment to take stock and pause a little. Just as beginning one's bleed is a rite of passage, so is ending it.

Many of the women I spoke to for their thoughts on final blood quickly mentioned their mothers. It seems like the end of one's cycle brings out the desire to talk to one's mother about her journey into the menopausal jungle. As women, we are spectacularly mismatched timing wise in this respect. When you start your period, often the last person you want to talk to about it is your mother. But by the time you end your periods, the person you most want to share the experience with is your mother. But often, she isn't around any more, or in that place where she can recall exactly how those hot flushes felt.

I dimly remember in my early twenties, when I was incredibly wrapped up in my own, new post-uni working life, my mother telling me she felt hot, odd and all over the place. Did I want to hear it? Not really. Did I take it in? Nope. Was it mildly annoying when we had to stop in the shops for a breather while she caught her breath and dealt with a wave of heat coming over her body? Yup. But you can bet your bottom dollar when I am facing down the barrel of the menopausal gun, I would love to hear those stories, her coping strategies and what she went through, as a way of navigating what's to come and bonding as women. I pray she is still around for many reasons, but also so I can selfishly tune in properly and learn about this stage.

My mum's mother died of cancer when she was only 61. She did have a hysterectomy – prematurely bringing her period career to a close – but my mum still craved her company and guidance when she went through 'the change'. It seems it's part of the cycle: wanting one's mother at the end of periods as you figure out the road beyond.

And life on the other side is just fine. Great even. I am reliably told.

As Louise puts it: 'When you are post-menopausal, in some ways you are back again. You are you. During the menopause, as your periods end, you can be teary, tetchy, forgetful. You can't do stuff you normally can. It can be very frustrating. But post menopause, it's all usually OK again. We are pretty damn cheery and that's not spoken about enough. We just don't have the same ups and downs anymore.'

The menopause itself is another huge taboo, which is slowly being demystified and thankfully broken down. And older women are finally becoming less invisible – although we have the furthest to go with that.

The point is, there is now a long post-period life for women. Statistically, we may be getting our periods earlier – but we have much more life on the other side of them too, during which we hope no longer to be devalued because our youth and ability to procreate has ended.

Miranda may loathe the flatness she feels as she enters this life chapter, but she also professes a love for menopausal women: 'I like women on the other side of periods. Menopausal women. They are tough. They are doing what they want. They are reinventing themselves and shrugging off all of their previous shit.' I definitely like the sound of that, even if it's a heck of a hormonal journey to get there.

Let's be clear. None of the women I spoke to about the end of their periods missed the blood – even the most passionate bleeders. For that sentiment, I have to turn to the words of someone described lovingly by her fans as 'the poet laureate of periods': Sharon Olds. The American writer and academic, now in her mid-seventies has written extensively about periods and the female experience in most of its forms. In 'When it Comes', Sharon sums up an awe and majesty about periods I haven't seen as exquisitely put anywhere else. If you have the chance, I recommend reading.

The factory inside of women is kind of amazing when you take a moment to consider what the hell is going on in there and the power of what it can do.

The period is a form of communication memo with a woman. It's a clear sign that this cycle is over and a child isn't in situ, that the new month is beginning and we're off again. It is like the body sighing and, in some people's case, groaning and dragging itself into the new month. Or roaring its way in there like 'Hello – let's rip this shit up!'

Without us realising it, our periods, however we choose to have them – chemically or naturally – do provide a rhythm; a heartbeat to the month. They also can dictate our moods, reactions and energy levels. Contrary to society's lampooning of women as moody hormonal cows during their periods, women miss them for these ups and downs. They can provide a pace and a variety we don't notice until they are gone.

Periods can also make women feel horny like no other time or synthetically induced moment. Lovers should take note. The hottest sex could be newly yours.

They can also create a momentum and a monthly release. And if you take it, the chance to reset.

By sharing the reasons why women miss their periods when they are gone, I hope you too can be more awake to the possible benefits your cycle gives you. You may never have noticed these plus points before, but now you actively can try to harness the good they bring. And, crucially, surrender yourself guilt-free at the points in your cycle you just feel like shit, safe in the knowledge that it's normal for you and you will rise again like a period queen from your squished-in sofa – ready to boss it.

I also aim for this knowledge to make us all a little more sensitive about what this change of period status might mean for someone who never had children, or who has lost a child. Women around you could be going through something pretty rough and now you may have an inkling why. I know I certainly do. Even if there's nothing I need to actively do with that knowledge – other than have it and be quietly respectful about the overwhelming power of the end of periods.

Personally, I find it riveting that we live so much longer than our periods nowadays; that there is so much life on the other side and that that life is just as fulfilling, if not more, than the teenage one we embarked at the start of our menstrual career. Thankfully, the days of only living as long as our periods are over – as are the presumptions that women are washed up and no longer useful to society once our perceived fertility is kaput.

Post periods, some women are only just getting started. It might be that periods gave women energy, ups and downs, healthy libidos and the ability to sprog, but once they are gone, women are freed from society's expectations, and some responsibilities. All bets are off. *How cool is that*? And that's a period-free future worth looking forward to – while respecting what the monthly blood gave you.

I will still be throwing that wild end-of-period party, whenever the moment comes, but for as long as I have to contend with my monthly foe, I am going to try and learn more about the benefits of my cycle (through gritted teeth), aided by fresh ginger, painkillers and greasy chips. I may not mourn my period's passing but playing it better by comprehending my cycle more is smart and I urge you and those you love in your life to do the same.

So my bleeding friends, we will get there, and I'll see you on the other side – if you don't beat me to it.

I'll be the one glugging a Bloody Mary with a suspicious looking stirrer...

LAST BLOOD

'Menstruation is the only blood that is not born from violence, yet it's the one that disgusts you the most.'

Maia Schwartz, an artist

Well done. Congratulations. Back slaps all round my friends. You have done it. We have done it.

You. Have. Read. An. Entire. Book. About. Periods.

And we're OK! We're more than OK. We are positively zinging my friends. Buzzing to get out there and tear this taboo up. Right?

I can confirm no wine was spoiled in the process. No dogs sent crazy. No teeth broken. No livestock harmed. No fields of crop destroyed with such flagrant and detailed descriptions of periods – the bodily fluid formally known as 'women's crazy juice'.

Dearest Reader, my aim was to instil within you a sense of period pride, whether you bleed or you don't, whether you love or loathe your flow, and especially if you have never really thought about menstruation in any great depth.

I hope I've achieved that by pointing out to you where some of the bizarre myths about periods come from, and who still believes and propagates them. I hope I've shown you the ludicrous lengths women still go to in order to hide their periods from general society and opened your eyes to some of the crazier things that have been allowed to be normalised or banned – all in the name of women's monthly blood.

We have met a heck of a lot of people along this merry, mad way…

From the politicians too embarrassed to utter non-threatening words such as tampon and vagina when discussing periods, to the trailblazers trying to make systematic and cultural change, such as US congressman Sean Maloney and UK MP Paula Sherriff.

From the women too poor to buy sanitary towels, like Rachel Krengel, to those fighting their corner, such as campaigner Amika George and teacher Chella Quint.

From the men who still cannot handle the biological fact that women bleed – like the forty-fifth President of the United States, a certain Donald J Trump – and use it as a way to ridicule us, through to those men doing all they can to understand menstruation, like comedian and single father Gary Meikle. Not forgetting Edgar Momplaisir, who simulated having a period for a week (now that's dedication).

From the sportswomen on the global stage finally

opening up about the impact of periods on their performance, like Chinese swimmer Fu Yuanhui and British runner Jessica Judd, to those trying out new approaches in the workplace, such as Bex Baxter, pioneering menstrual leave in the UK.

From those who desperately crave periods like Jen Palmer to those who would do anything to make theirs stop, such as Cass Bliss, the Period Prince.

From those who really, really love the red stuff between the sheets, a largely anonymous pack of bloodhounds, to those who detest 'the taste of raw steak' and the mere idea of period sex turns them tampon white.

From the business-folk and advertising executives stifling the progress of new players in the period product space, to those trying to disrupt a stubborn industry with new blood absorption technology and renewable solutions at Thinx and Dame.

From the religious and superstitious folk still banning women from temples and farms, to those fighting back with powerful words and imagery, such as the poet Rupi Kaur.

It's been a pretty wild trip we've been on together, revealing stories and insights which will live long in my memory and, fingers crossed, yours too.

Hopefully you haven't curdled any mayonnaise simply by touching this book – or, you know, ended up with a tampon up the tush.

Hopefully you've had a wry laugh along the way – especially at the thought of embarrassed women squirrelling away their period sex sheets, dislodging stuck tampons before business meetings, and upon learning that one of the world's richest companies, Apple, simply forgot about periods. *Forgot*.

You may have even felt tearful. I know I did on more than one occasion, reflecting on the dramatic moment Jen Palmer was told she was born without a womb, reading Lena Dunham's recollections about having hers removed, discovering the anecdotal stories of those girls in Britain skipping school because they are too poor to afford tampons, and learning how sanitary products are so cruelly used against homeless women.

One of my other humble hopes was to point out things you may never have noticed or known before: why tampons are really taxed, how periods can land women with criminal records, the silly battle for a period emoji, the bitter fight for the decent sex education of girls and boys, and how one government is paving the red brick road to having a period utterly free of charge. Tampons and pads galore! Oh bonny, wise Scotland, we salute your valiant endeavours on behalf of womankind.

And, apart from making your jaw drop at times – in both good and bad ways – I hope to have ushered some new perspectives into your view.

A lot of people (men, women, family, friends and strangers) who asked me what I was writing a book about recoiled with a mixture of confusion and nervous displeasure when I replied: 'periods'. Some asked, did I mean *those* periods. The ones in the knickers? As opposed to periods of time. Once I duly confirmed that I did indeed mean 'those' periods, they then often followed up with: 'But what on earth is there to say about them?'

Slightly tired of justifying or explaining my topic of choice, I found myself smiling and replying: 'Just buy the book and see? And do come to the launch party where we'll be drinking Bloody Marys stirred with tampons.' The party idea definitely made the subject seem more palatable – even if a soggy tampon proved a jape too far.

But the truth is, I have never wanted to write a book before. I genuinely hadn't. Until this topic kept following me around, tapping me on the shoulder, looking at me with puppy dog eyes and making me realise there was a need to bust the myths around periods and change the way women are viewed in light of them. And in turn, alter the way women perceive themselves.

As it turns out, there is a shy hunger to talk about periods and what they mean to different people. There are also brilliant, funny and touching tales that emerge when the period taboo is prodded, peeled away and people feel happy to proceed. Periods are a great untold part of women's stories and they feed into every part of our lives: health, body, identity, mood, personality, mindset, sex lives, fertility, a girl's coming of age and menopause.

It's at this point I wish to stress something important.

Some women don't experience any mood changes or even pain (lucky bastards regarding the latter, I envy you in the extreme) when they get their period – but even if they do, it does not mean they are no longer capable. That is not the point. If anything, having a period which does alter your mood or makes you feel generally a bit crap, and then continuing to do everything as you were before, makes you *more* capable. Not less.

You personally may have preferred not to have completed that project or given that speech on that particular day, but most women still function, and function well, while bleeding. Remember and repeat after me: Paula Radcliffe broke a world record for running a marathon while fighting period cramps in the last third of the race. Hell yeah.

Equally, there should be no shame in giving women the space to talk about it when they do feel off, and empowering them to make suggestions for what could improve their day: a hot water bottle at their desk, an open window or a gargantuan biscuit. And yes, of course some women might take the piss and use their periods to skive or shirk their duties. There are always a few. People misuse sick leave to cope with nightmare hangovers but we don't ban sick leave. But the vast majority of women would benefit from being the real, honest version of themselves both at work and at home.

The point is, until periods stop being a brush to tar women as weaker, less capable and overly emotional beings, we cannot move forward to being seen as full equals. Instead, we remain separate, dirty and weird. Less than men. Ashamed.

One high profile, older journalist, whose anonymity I shall of course preserve, who I approached on email to tell me about her relationship with periods, replied in a memorable fashion: 'I'm sorry but my own period remains personal to me and I don't want to share.'

Her silence is her right. She may also have had something very personal happen to her regarding her period. Or not. Perhaps everything was tickety-boo and she simply wished to keep her stories to herself. But somehow I felt chastened by her response. Like I had asked an inappropriate question.

And that's the thing. Periods have been the unmentionable for too long – rendering them exactly what they shouldn't be: mysterious, weird and disgusting things that happen in the darkness of our knickers.

It's time to shine the light on our stained underwear and stop being private about one of the processes we all owe our lives to. Those things which continue to live in the dark usually divide us unnecessarily and always benefit from the healthy disinfectant of sunlight.

The real dream? For periods to become truly unremarkable. Not to be worthy of laughter or piss-taking by men of all ages. Not to make women seem like a weak, foreign or filthy species. Not to make women feel embarrassed in front of each other.

That's what our red quest is ultimately aiming for: a new understanding, respect and, essentially equality.

To aid us in our efforts to reach that point, I've compiled a list of concrete actions you and your friends could take upon yourself to help us get there. Introducing the jazzily entitled 'Period Pride Manifesto', for us to keep together...

PERIOD PRIDE MANIFESTO

- No more lying about periods. If you have one and need to mention it, *do*. And without turning period pink.
- No more listening to period bullshit – religious or otherwise. (Banter included.)
- Have the courage to call out all said bullshit. Humour is always your friend during confrontations, but sometimes what you are dealing with ain't funny.
- Ask for anything you need to make periods easier, at school and at work. Educate those around you in doing so.
- Be a good friend, colleague, boss to those having periods who are starting to talk about them.
- Donate tampons and pads wherever you can: supermarkets, school toilets, homeless shelters, food banks.
- Petition your MP for period products to be free to all, like loo roll.

- If you feel horny on your period – have sex. Without shame. Don't just give out oral sex because you feel you owe some weird sex debt.
- Make and star in your own truthful period advert, if you wish. And share it online.
- Observe your own cycle and note down the positives (if there are any) and particulars to make life easier each month.
- Do your research and experiment with new period products like you would with recipes or sportswear.
- Go to the doctors if you don't feel right on your period or during sex or anything else gynaecologically. And don't take no for an answer if you feel fobbed off. Push to see specialists and fight for solutions to improve your lot. Medicine should be your friend and your body shouldn't be ignored.
- Touch mayonnaise while menstruating. A lot of it. Much to the surprise of anyone else still believing bleeding women spoil mayo (most of Madagascar it seems, by the way), it won't curdle.

Our task is to make menstruation so totally and utterly unremarkable for the next generation that when the word 'period' is mentioned no eyebrows are raised in disdain or disbelief.

To be clear, this does not mean erasing periods. Or doing down women's experience of them, which can be brutal. It means elevating them from the dark underworld of our knickers to normality – so that men, women and children are no longer appalled by periods.

And so women and girls can talk freely about them. To borrow the quote of Edgar Momplaisir, the only man I interviewed who tried 'menstruating': 'Women need to be able to talk about their periods as openly as guys talk about getting wood.' Charming, but it makes the point perfectly.

The saddest sight I saw during the research for this book? The videos of Hindu women protesting at the lifting of the menstrual ban at the holy Sabarimala temple in India last year (which I mentioned earlier in the book). These women were attacking and trying to stop – alongside many, many angry men – two brave women attempting to head inside. It's worth noting that this plucky duo still wanted to go inside, despite the temple's chief, Prayar Gopalakrishnan, only two years prior, suggesting he would only allow women entry if a machine was invented to detect they were 'pure', meaning they weren't menstruating. This is not normal or acceptable.

Women themselves believing they are too impure to enter a religious space because of a natural bodily process is seriously messed up. Especially when you consider that, in Hinduism, women are considered literally divine. Worshipping the female aspect of God is a key tenet of Hindu philosophy. So, it would naturally follow, as one female Indian activist told the BBC, that 'if a woman is divine then her menstruation is divine'. Quite.

Remember team, period pride doesn't mean loving your period. I never will. In fact, I ignored mine for far too long when it was trying to tell me something serious about my body. But it does mean destroying one of the final and most pernicious taboos about being a woman.

The former and first female US Secretary of State Madeleine Albright once said: 'It took me quite a long time to develop a voice, and now that I have it, I am not going to be silent.'

We may have thought we'd said it all about being a woman. But we haven't. Not even close. The period taboo must die a death and with it all the negative nonsense still believed about women and by women, rendering them less than men.

You cannot legislate a taboo out of existence. Such a cultural shift has to come about through decisive actions, leading by example and sheer force of will – which women, menstruating or not, certainly aren't lacking.

You know it. I know it. Together, *we* know it and can make the change happen. Because guess what?

It's about bloody time. *Period.*

RIDING THE COTTON UNICORN:

A handy appendix of period euphemisms

Who knew? I mean seriously, who the hell knew? Apparently, there are more than 150 ways to say you are on your period without actually *saying* the word period.

Of course, the very best word for period, is you know, period. And while this handy appendix I've compiled will undoubtedly make you giggle, pull a face and perhaps wince like you've eaten an extremely sour sweet from the nineties, we do need to be able to do away with these childish pseudonyms and call a period, a period. And fast. A dark sense of humour is often necessary during one's period so hopefully this list will also prompt you to laugh (or snort derisively at some of the euphemisms men have blatantly come up with).

Such a vast dictionary of 'acceptable' and ludicrous alternatives is almost the best proof of how culturally taboo periods stubbornly remain. No one's blushes need be spared when discussing a monthly biological occurrence. And yet the fantastical lengths people have gone to in order to describe menstruating – when there is a perfectly suitable and non-gory word for it already in circulation – is pure madness.

So, dear Reader, I share this appendix as evidence to you, the jury; evidence of how much it is bloody time to call a period by its name, and no other.

Below are some of the most common period pseudonyms in the English language and, for good measure, I've also done a mini trip around the world and included some of those weird and wonderful phrases at the end.

Despite its vastness, this list is by no means exhaustive. For contributions, I must thank my very creative friends, none of my family (who didn't understand why one would need a euphemism), some anonymous internet folk and the people behind the Wear White Again heavy period campaign who commissioned some research in this space.

My personal favourites are 'shark week', 'riding the cotton unicorn' and the old-school French take, in which the arrival of a period is genuinely referred to as: *les Anglais ont débarqué* – 'the English have landed'. In our redcoats (the former name for the English army during Napoleon's time). Old habits die hard.

I have taken the liberty of grouping the list into themes, from food-based euphemisms through to phrases men have come up with to denote their sexual starvation during women's periods.

Warning: food-related terms may put the weak-stomached off certain ingredients for a while. I haven't had a jam sandwich since I was a kid, but if I ever indulge again, it won't quite have the same sweet taste. Meanwhile, nothing could ever put me off chopped liver.

Politically Russians, Communists and the Red Army feature heavily. As do vampires, the devil and Dracula.

Personally, I feel sorry for all the grannies and aunties out there. And anyone called Tom, Rita, Irma, Chico, George or, ahem, Emma. Your names are regularly being used to describe someone's uterus lining shedding into their knickers.

Deep breath, here we go:

Food
Bloody Mary
Cherryade is here
Chopped liver
Dracula's sandwich
Having ketchup with your steak
I have got a sloppy burger
Ketchup
Ribena
Sinner's juice
Strawberry week
Tomato soup

Blood
Bleeding beauty
Bleeding clam season

Bloody Niagara Falls
Blood rain
Bloody Sundays
My bleeds

Visits
A friend is visiting
George is here
Got my Emma
Got my enemy
Got my Mary
Granny is visiting
Monthly friend
My friend Tom
Old friend
Special friend
Venus is visiting

Aunties
Auntie Flo's in town
Auntie's here
Aunt Irma is visiting
Auntie is visiting from Argentina
Aunt Rita is visiting
Aunt Ruby is visiting

Male slang
Being on a red traffic light
Blow-job week
Gravy

Mistress time
Out of commission
Up on blocks (like a car at the garage, aka out of service)
Up on bricks

Jam
Jam butty
Jam pasty
Jam sandwich
Jam time
On jam bag

Red
Checking into the red roof inn
Code red
Code rouge
Driving through redwood forest
I've got the reds
In the red
Little red mouse week
Off visiting the red planet
On your redder
Painting the town red
Red Army's in town
Red badge of courage
Red cat
Red dessert
Red dragon
Red flag is flying
Red Indian visit

Red letter day
Red rag
Red river
Red roses
Red Sea
Red wedding (à la *Game of Thrones*)
Reds are playing at home
Riding the red tide
The red baron is visiting

Miscellaneous
A visit from Captain Bloodsnatch
Arsenal/Liverpool playing at home
Bad week
Builders are in
Bunny week
Cardinal has his hat on
Coming on
Crime scene
Devil's waterfall
Feeling periodical
Female moment
Funny fanny
Girl flu
Granny gagger
Granny pants week
Having a baby stopper
Having your Nellie
In dry dock
Lady business

Luna phase
Mickey Mouse is in his clubhouse
Moon time
Mother nature's gift
Mother nature is in town
Mr Blobby's in town
My menses
On fire
On heat
On the blob
Pad time
Painters are in
Plug time
Potato harvest
Punched in tummy
Rag week
Red tent time
Riding the cotton unicorn
Rollover week
Russian Army
Saddling up old rusty
Shark week
Star week
Surfing the crimson wave
Tearing down wallpaper time
That time of the month
The Communists have invaded the summer house
The curse
The dread
Trip to blood mountain

Vampire sandwich
Vampire's packed lunch
Walk like an Egyptian
War
Women's fire
Women's time

Around the world
America: the Red Sox have a home game
Australia: I've got the flags out
Brazil: I'm with Chico
China: little sister has come
Denmark: there are Communists in funhouse
Ethiopia: the monthly flower
Finland: mad cow disease (charming name for PMS)
France: the English have landed
Germany: the cranberry woman is coming
Ireland: I am wearing a jam rag
Japan: Little Miss Strawberry
Latin America: Jenny has a red dress
Netherlands: the tomato soup is overcooked
Puerto Rico: did the rooster already sing?
Romania: Santa Claus has come
South Africa: Granny's stuck in traffic
Sweden: lingonberry week
Turkey: the motherland is bleeding

Any more for any more? I am all ears.

In the meantime, if you've never called a period, a period, why not try it? Go on, I dare you. I'm sure all the grannies and Auntie Ritas out there will agree with me.

A LONG OVERDUE LETTER TO MY PERIOD

(Written while cramping)

I am standing in a draughty picturesque church on an icy Saturday in December wearing four-inch gold glittery heels silently cursing you, as you churn my insides up without a thought for the occasion. Try as I might to focus on my friend, the gleamingly beautiful bride, and stand stoic during the hymns she's painstakingly chosen for us to chorus, all I can think is: will you please fuck off? And when the fuck can I rip off this fascinator, my heels, dress, tights and, while I am at it, my skin? The need for delicious wedding champagne to dull your grip is getting increasingly urgent. Yeah, I'm talking to you. My pushy, aggressive, attention-seeking period. The very same period that has no sense of timing or mercy. OK, fine – due to the wonder of the pill you do arrive on time, mostly. But you pay no heed as to what I am doing and how you transform me from a fast walking and talking, vibrant being into a husk who craves warmth, trousers which have lost their elastic and copious amounts of fat chippy chips doused in vinegar.

I thought it high-time I addressed you directly – seeing as we've known each other intimately for twenty-four years, which, bar my parents and three long-suffering school mates from Manchester, makes you my longest relationship. And yet the dialogue has been mainly one way. You communicate with me loud and proud – while I submit.

I feel you coming, long before you officially arrive. And yes, I've got a period disease, which makes you more brutal to me than most women's menstrual partners. But even when I've got you under some form of control via the pill and a heady cocktail of painkillers, you are still a law unto yourself, and therefore me.

You make me feel hot and cold. Blue and bluer. I feel your downward pull so strongly it makes my legs go weak and gives gravity a run for its money. You wreak havoc with my bowels, skin, mood and energy levels.

In a way, if you weren't so bloody and painful, you would be my kinda person: loud, dramatic, uncompromising, definite and game-changing. If you weren't so busy making me feel like a crock of useless shit, I'd go drinking with you.

Except, you'd be like that mate who always takes it one step too far. The one you have to keep your eye on. Everyone's got one. That pal who sinks more booze than they can handle, requires your undivided attention and sours the mood at even the happiest of gatherings.

But you must be having a right old laugh at humankind. There you are with your comrades, every month, bold as brass, unapologetically coming to do your disruptive thing to me and my sisters and we don't breathe a fucking word. You are loud and we are cowed.

Do I tell a soul at my girlfriend's wedding why I've got a face etched with pain, and mainlining booze almost as fast as I'm hoovering up canapés (stationed outside the kitchen) and why I'm hobbling in my heels long before the band starts? Do I hell. You demand my attention and refuse to be ignored. Women everywhere could learn a lot from your boldness. But a deeply-embedded social code has developed around you which commands women's silence and shame.

Having given it some serious thought, I can't call ours a 'love–hate' relationship. I can't forgive how you rob me of me each month. What I can stretch to is 'respect–hate'.

The three highlights of our union? Our first meeting and my sheer relief that you showed up and made me feel like a 'normal' teenage girl – whatever that is. I silently and naively thanked you for putting in an appearance in that nippy loo in Manchester's House of Fraser. Our second peak revolved around school sport. I must salute you for all the times you got me out of dreaded swimming lessons – whether you were there or not. But our best moment? When you finally buggered off post IVF. That's when I truly respected you for the first time and understood your purpose: our son.

For without you, or at least some semblance of you, I wouldn't have him. Other women will feel utterly different towards their period. It will mean something else to them entirely. But at that point, you at last made sense to me and my gratitude was enormous. Somehow, in our bloody war, we had done something. Together.

I will still have to work hard to tame you. I know I will continue to loathe the equipment you force me to shove into my pants to cope with your presence. I will still damn each day you arrive and the impact you have. And I know I will zealously celebrate the day you swagger off for good – despite the associated emotions of getting older clamouring for my attention.

But we've been through a lot. A heck of a lot. And we could have at least another decade or two together.

So, if you are going to stay loud and proud then so the hell am I.

Yours, in grudging, trudging admiration,

Emma

STRAWBERRY WEEK. THE REDS. MY MENSES. RED WEDDING. M
BLOODSNATCH. DRACULA'S SANDWICH. OLD FRIEND. OUT OF
MONTH. BLOODY NIAGARA FALLS. GOT MY MARY. FEMALE MON
RED. TRIP TO BLOOD MOUNTAIN. VENUS IS VISITING. THE CURS
WEEK. RED FLAG IS FLYING. WALK LIKE AN EGYPTIAN. BLOODY
HARVEST. HAVING A BABY STOPPER. SHARK WEEK. ON HEA
WATERFALL. JAM PASTY. RED ROSES. GRANNY GAGGER. FEEL
AUNTIE FLO. RUSSIAN ARMY. TEARING DOWN WALLPAPER TIM
MOON TIME. TOMATO SOUP. CRIME SCENE. GIRL FLU. THE RED
BLOB. BLEEDING BEAUTY. BLEEDING CLAM SEASON. HAVING
WITH YOUR STEAK. PUNCHED IN TUMMY. UP ON BRICKS. SLOPP
GIRL FLU. JAM SANDWICH. VAMPIRE'S PACKED LUNCH. RED E
RED TENT TIME. MONTHLY FRIEND. LADY BUSINESS. RED LETT
BLOODY MARY. CHOPPED LIVER. TEARING DOWN WALLPAPER T
KETCHUP WITH YOUR STEAK. RED LETTER DAY. JAM PASTY. S
WEEK. RED ROSES. TOMATO SOUP. PLUG TIME. WALK LIKE AN
NIAGARA FALLS. HAVING YOUR NELLIE. BLOOD RAIN. RED BAD
PUNCHED IN TUMMY. IN THE RED. VAMPIRE SANDWICH. THE DR
FRIEND. CRIME SCENE. THE CURSE. BLOODY SUNDAYS. OLD F
SHARK WEEK. RED LETTER DAY. AUNTIE FLO. RED TRAFFIC LIGH
BUSINESS. GIRL FLU. UP ON BRICKS. GRANNY GAGGER. JAM BU
RED MOUSE WEEK. VISITING THE RED PLANET. WOMEN'S FIR
THE RED TIDE. STAR WEEK. RED WEDDING. THE RED BARON IS
TOWN RED. CODE ROUGE. BUILDERS ARE IN. COMING ON. DE
HAVING A BABY STOPPER. LUNA PHASE. RED RIVER. MOTHER
PAINTERS ARE IN. POTATO HARVEST. RAG WEEK. RIDING THE C
RED TIDE. SURFING THE CRIMSON WAVE. JAM SANDWICH. VAM